American Oil Diplomacy in the Persian Gulf and the Caspian Sea

Florida A&M University, Tallahassee
Florida Atlantic University, Boca Raton
Florida Gulf Coast University, Ft. Myers
Florida International University, Miami
Florida State University, Tallahassee
University of Central Florida, Orlando
University of Florida, Gainesville
University of North Florida, Jacksonville
University of South Florida, Tampa
University of West Florida, Pensacola

American Oil Diplomacy in the Persian Gulf and the Caspian Sea

Gawdat Bahgat

University Press of Florida
Gainesville/Tallahassee/Tampa/Boca Raton
Pensacola/Orlando/Miami/Jacksonville/Ft. Myers

Copyright 2003 by Gawdat Bahgat
Printed in the United States of America on acid-free paper
All rights reserved

08 07 06 05 04 03   6 5 4 3 2 1

Library of Congress Cataloging-in-Publication Data
Bahgat, Gawdat.
American oil diplomacy in the Persian Gulf and the Caspian Sea /
Gawdat Bahgat.
p. cm.
Includes bibliographical references and index.
ISBN 0-8130-2639-3 (cloth: alk. papaer)
1. Petroleum industry and trade—Political aspects—Persian Gulf
Region. 2. Petroleum industry and trade—Political aspects—Caspian
Sea Region. 3. United States—Foreign relations—Persian Gulf Region.
4. Persian Gulf Region—Foreign relations—United States. 5. United
States—Foreign relations—Caspian Sea Region. 6. Caspian Sea Region—
Foreign relations—United States. 7. Geopolitics—Persian Gulf Region.
8. Geopolitics—Caspian Sea Region. 9. World politics—21st century.
I. Title.
HD9576.P52B34 2003
333.8'23'0973—dc21   2003040238

The University Press of Florida is the scholarly publishing agency
for the State University System of Florida, comprising Florida A&M
University, Florida Atlantic University, Florida Gulf Coast University,
Florida International University, Florida State University, University
of Central Florida, University of Florida, University of North Florida,
University of South Florida, and University of West Florida.

University Press of Florida
15 Northwest 15th Street
Gainesville, FL 32611–2079
http://www.upf.com

Contents

List of Tables   vii

List of Maps   ix

Preface   xi

List of Abbreviations   xv

1. U.S. Energy Security   1
2. The Global Energy Scene   24
3. Managing Dependence: American-Saudi Oil Diplomacy   43
4. The United States and Iraq: Continuity and Change   72
5. The United States and Iran: Prospects for Rapprochement   103
6. The Geopolitics of the Caspian Sea   140

Glossary   175

Notes   179

Bibliography   193

Index   207

Tables

1.1. U.S. fossil fuel consumption, 1990–2020   5

1.2. U.S. oil production and consumption, 1970–2000   7

1.3. Imports from the Persian Gulf region as a percentage of net oil imports, 1983–2000   21

2.1. Fossil fuels in the United States and the European Union, 2000   27

2.2. Fossil fuels in Pacific Asia in 2000   35

3.1. Crude prices in $U.S. per barrel, 1972–2000   48

3.2. Saudi oil production and share of total OPEC, 1970–2000   52

4.1. Iraq's oil production, 1970–2000   77

5.1. Iran's oil production, 1970–2000   110

6.1. British Petroleum/Amoco's estimate of the Caspian's resources, 2000   144

6.2. *Oil and Gas Journal*'s estimate of the Caspian's resources, 2000   144

6.3. U.S. Energy Department's estimate of the Caspian's resources, 2000   144

Maps

1. The Arctic National Wildlife Refuge, Alaska   3
2. The North Sea and northwestern Europe   25
3. Saudi Arabia   44
4. Iraq   73
5. Iran   104
6. The Caspian Sea   141

Preface

In this book, I do not claim to analyze American foreign policy in the Middle East. Similarly, I make no attempt to examine U.S. energy strategy comprehensively. Rather, I focus on Washington's efforts to create and maintain a state of "energy security." Specifically, I provide a thorough analysis of U.S. relations with two energy-rich regions, the Persian Gulf and the Caspian Sea.

The first chapter outlines the main themes of the study. It argues that a long-term American energy strategy should include a broad combination of measures that would stimulate domestic production, provide incentives for conservation, promote clean technologies, and eliminate political barriers to world markets. Furthermore, despite the increasing supplies of oil and natural gas from the North Sea, Latin America, and southern and western Africa, and the projected contribution of the Caspian Basin to global oil security, the Persian Gulf remains the main reservoir that can meet either a substantive increase in demand or an emergency caused by a major disruption of supplies.

Chapter 2 argues that U.S. energy policy cannot be understood in isolation from that of other countries. The focus is on the global energy scene. Specifically, the energy strategies of the European Union, Russia, and Pacific Asia are examined. In addition, the controversy regarding global warming and the Kyoto Protocol is analyzed. The conclusion is that, given the globalization of energy markets, there is a need to adopt a collective approach to energy security. A narrow national approach by an individual state will not succeed.

Chapter 3 highlights the mutual understanding and close cooperation between the United States, the world's largest oil consumer and importer, and Saudi Arabia, the world's biggest producer and exporter. The analysis focuses on two dimensions of American-Saudi energy relations: the stability of oil prices and the gradual reopening of the Saudi hydrocarbon sector to foreign investment. Also discussed are some of the main challenges to this long-term partnership between Washington and Riyadh. These challenges include security, economic and political reforms, the Arab-Israeli conflict and peace process, and militant Islam.

Since the demise of the monarchy and the establishment of a republican system in Baghdad in 1958, tension, suspicion, and outright hostility have characterized American-Iraqi relations. Iraq, which holds approximately 10 percent of world oil proven reserves, has been dissatisfied with its geographical makeup. Territorial disputes are common in the Persian Gulf, but more than the other regional states, Iraq has sought to realize its claims aggressively and militarily. With substantial strategic and economic interests in the Gulf region, the United States has taken the lead in containing the perceived Iraqi threat and promoting stability in both the Gulf states and the global oil markets. Developments since the September 11 terrorist attacks have intensified the debate in Washington regarding the next step with Baghdad. Chapter 4 suggests that the Iraqi dossier is far from being closed. Washington has yet to find a way to reconcile the need to keep Iraqi oil flowing into the global markets with its desire to deter Baghdad from committing any aggression against its neighbors or acquiring the means to pursue such a policy.

The troubled American-Iranian relations since 1979 are the subject of chapter 5. Iran can be seen as an energy giant with one foot in the Persian Gulf and the other in the Caspian Sea. Still, the Islamic Republic's oil and gas potentials have been restrained by its tense relations with the United States and by Washington's efforts to deny Tehran access to badly needed foreign investment to update and modernize its energy infrastructure. The chapter discusses the American sanctions against Iran and the main obstacles toward a rapprochement between the two countries. These include sponsoring terrorism, acquiring and developing weapons of mass destruction, and opposing the Arab-Israeli peace process. Despite the huge gap between Washington and Tehran on these three issues, the two sides share mutual interests and concerns in several other areas such as Afghanistan, Iraq, and global energy markets.

Since the early 1990s, many energy experts and policymakers have perceived the Caspian Sea as one of the most promising oil and gas reservoirs in the world. Experience suggests that this euphoria has been exaggerated. Several logistical, economic, and political obstacles need to be addressed and overcome before the region can realize its projected potentials. The final chapter examines six of these challenges: an accurate assessment of the Caspian Sea's hydrocarbon resources; rivalries between regional and international powers; domestic ethnic divisions; disputes over the legal status of the Caspian; disagreements over the most cost-effective transportation routes; and the potential geopolitical changes in the region in the aftermath of the war on terrorism.

In short, this volume examines the growing dependence of the United States on fossil fuels, particularly oil, and the main challenges it faces in securing supplies from the Persian Gulf and the Caspian Sea. The study will be useful to students, scholars, and individuals interested in international relations, international economy, Middle Eastern policy, and energy studies.

Many people helped me in writing this book. I would like to thank Sandra and Pat Dickson, Beth and Steven Sims, Gene and Helen Hooker, Anthony McDermott, and Michele Reynolds-Jackson. Without their inspiration, this work would not have been completed.

Most of the research and writing of this volume were completed in 2001, and the book was updated in light of the developments that followed the September 11 terrorist attacks in the United States. In March 2003 an international coalition, led by the United States, started a war with Iraq aimed at destroying that country's weapons of mass destruction and ousting Saddam Hussein's regime from power. The ramifications of this new conflict are uncertain and will need some time to be understood. By the time this book went to press, the dust has not yet settled in this new war in Iraq, and the 2003 war is not covered.

Abbreviations

AIOC	Azerbaijan International Operating Company
ANWR	Arctic National Wildlife Refuge
Aramco	Arabian American Oil Company
bcm	billion cubic meters
b/d	barrels per day
BP	British Petroleum
CASOC	Californian-Arabian Standard Oil Company
CIA	Central Intelligence Agency
CIS	Commonwealth of Independent States
COP	Conference of the Parties
CPC	Caspian Pipeline Consortium
CWC	Chemical Weapons Convention
EIA	Energy Information Administration
EPCA	Energy Policy and Conservation Act
EU	European Union
GDP	gross domestic product
IEA	International Energy Agency
IGAT	Iran gas trunk line
ILSA	Iran-Libya Sanctions Act
INA	Iraqi National Accord
INC	Iraqi National Congress
INOC	Iraq National Oil Company
IOCs	international oil companies
IPC	Iraqi Petroleum Company
KDP	Kurdistan Democratic Party
LNG	liquefied natural gas
mcf	million cubic feet
MEK	Mujahidin-e Khalq'
MGS	Master Gas System
mpg	miles per gallon
NBC	nuclear, biological, and chemical weapons
NEPDG	National Energy Policy Development Group

NIOC	National Iranian Oil Company
OECD	Organization for Economic Cooperation and Development
OPEC	Organization of Petroleum Exporting Countries
PSA	production-sharing agreement
PUK	Patriotic Union of Kurdistan
SABIC	Saudi Arabian Basic Industries Corporation
SAIRI	Supreme Assembly for the Islamic Revolution in Iraq
SPR	Strategic Petroleum Reserve
tcf	trillion cubic feet
TCP	Trans-Caspian Pipeline
TPC	Turkish Petroleum Company
TSC	technical service contract
UAE	United Arab Emirates
UN	United Nations
UNFCCC	United Nations Framework Convention on Climate Change
UNSCOM	United Nations Special Commission
WMD	weapons of mass destruction
WTI	West Texas Intermediate
WTO	World Trade Organization

U.S. Energy Security

Since 2001, a top priority of George W. Bush's White House administration has been energy security. Both the president and his vice president, Dick Cheney, were involved in the oil industry before they were elected to office. In 1977, Bush founded Arbusto Energy, Inc., which was merged into Harken Energy Corporation in 1986. Cheney was the chief executive officer and chairman of Halliburton Company before joining the Bush ticket. Given their backgrounds and the perception of energy crisis, which has persisted since the late 1990s, Bush decided to address the question of the nation's energy needs shortly after taking office. In his second week in office, the president established the National Energy Policy Development Group (NEPDG), directing it to develop a national energy policy. The group, headed by Cheney, included the secretaries of state, treasury, interior, agriculture, commerce, transportation, and energy as well as the heads of the Federal Emergency Management Agency, the Environmental Protection Agency, and other top officials. In May 2001, the NEPDG presented its assessment of the "energy crisis" in the United States and a long list of recommendations to avert the crisis and enhance the nation's energy security. The findings and recommendations of the NEPDG have incited an intense debate among policymakers, professionals in the energy industry, journalists, academicians, and others on the current status and the future of the country's energy policy.

Putting the rhetoric aside on whom to blame for the current shortage of energy supplies, a consensus has emerged.

(1) The nation has not had a comprehensive, integrated, strategic energy policy for decades. Both Democratic and Republican administrations have failed to articulate such a policy. Instead, they have provided short-term solutions to evolving crises, which proved inadequate to address the long-term imbalance between supply and demand. Indeed, most federal energy initiatives were

in response to emergencies following large increases in petroleum prices but were allowed to lapse afterwards.
(2) Since the late 1940s, every American recession has been preceded by spikes in energy prices. The economic slowdown of the early 2000s is no exception.
(3) The demand for energy skyrocketed in the 1990s. This can be explained by the fact that the rise in energy consumption usually occurs during periods of relatively low or stable prices and during periods of healthy economic growth.
(4) This rising demand, however, was not matched by an expansion in supply. There was severe underinvestment in energy infrastructure partly due to low industry profits through much of the 1990s (as a result of low and stable energy prices). This underinvestment was exacerbated by tightened environmental restrictions.
(5) Finally, liberalization of energy policy has facilitated efficiency and the smooth allocation of resources. But rapid deregulations of the oil, natural gas, and power sectors have also reduced the incentives for specific businesses to invest in large inventories or excess capacity that can help smooth markets during times of disruption or unexpected volatility in demand growth.[1]

Solutions have been proposed to form a workable energy policy:

(1) The fundamental imbalance between supply and demand, if allowed to continue, will inevitably undermine the nation's economic growth, standard of living, and national security. Supply and demand need to be addressed simultaneously.
(2) Reviving the public's interest in energy efficiency and conservation at home is an important component of a future energy policy. Even with improved energy efficiency, however, the United States will need more energy supply.
(3) These needed energy supplies should be diversified in two senses: the mix of energy sources and the geographical origin of that energy.
(4) The demand for energy will most likely continue to outpace domestic production. As a result, the United States will increasingly rely on import from producers across the globe. A reassessment of the role of energy in American foreign policy is needed.

To sum up, a long-term energy policy would include a broad combination of measures that would stimulate domestic production, provide incentives for conservation, promote clean technologies, and eliminate po-

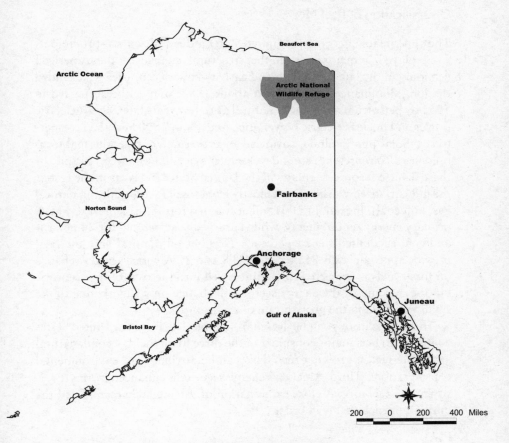

Map 1. The Arctic National Wildlife Refuge, Alaska

litical barriers to world markets. These issues will be examined in detail in this chapter. The analysis suggests that in spite of the need to diversify the fuel mix (oil, natural gas, coal, nuclear power, hydroelectricity, and other renewable resources), oil will continue to play a significant role in the nation's energy sector. Furthermore, the growing trade in energy with a variety of suppliers—North America, West Africa, the Caspian Basin—is important to enhance American energy security, but the nations of the Persian Gulf region—Bahrain, Iran, Iraq, Kuwait, Oman, Qatar, Saudi Arabia, and the United Arab Emirates—will continue to dominate the global energy market. In short, securing oil supplies from the Persian Gulf region will be a fundamental goal of American foreign policy.

Diversification of Fuel Mix

The appetite for energy as the United States industrialized was prodigious. Coal played a major role in the structural change of the American economy in the late nineteenth and early twentieth centuries. Coal ended the long dominance of fuel-wood about 1885, only to be surpassed in 1951 by petroleum and then by natural gas a few years later. Hydroelectric power and nuclear electric power appeared about 1890 and 1957, respectively.[2] Solar photovoltaic, advanced solar thermal, and geothermal technologies also represent recent developments in energy sources. In spite of these diverse resources, energy in the United States today, as in the rest of the industrialized world, comes mostly from fossil fuels (crude oil, natural gas, and coal). Indeed, in 2001, oil accounted for 38 percent of total U.S primary energy requirements, while natural gas accounted for 24 percent and coal accounted for 23 percent.[3] Together oil, natural gas, and coal provide approximately 85 percent of U.S. energy. America's heavy reliance on fossil fuels is projected to continue well into the twenty-first century, but the makeup of the energy sector will change only slightly due to the small variation in the growth rate of each source.

Three conclusions can be drawn from table 1.1. First, the United States continues to be a major consumer of the three fossil fuels. Second, natural gas is the fastest growing fuel. This can be attributed to environmental considerations. Third, America's energy security is enhanced by this diversity in the fuel mix. A close examination of the main characters and the outlook of each fuel is in order.

Coal

Coal is America's most abundant fuel source. The United States has the largest coal proven reserves in the world, enough to last for another 250 years at the current level of consumption. On the other hand, coal is one of the dirtiest sources of energy and consequently the least favored by policymakers and the general public. Coal combustion produces several types of emissions that adversely affect the environment. The combination of these two characteristics has influenced the production, uses, pricing, and outlook of coal. Since 1950, the United States has produced more coal than it has consumed. Technological improvements in mining and the shift toward more surface-mined coal, especially west of the Mississippi, have led to great improvements in coal mining productivity. This excess production has allowed the United States to become the fourth largest coal exporter in the world, behind Australia, South Africa, and Indonesia. The

Table 1.1. U.S. fossil fuel consumption, 1990–2020

Fuel	1990	1999	2005	2010	2015	2020	Average annual percentage change, 1990–2020
Oil	846	957	1,043	1,139	1,231	1,310	1.5
Natural gas	486	562	659	727	810	873	2.1
Coal	481	548	606	640	659	689	1.1

Note: All figures are in million tons of oil equivalent.
Source: Energy Information Administration, *Annual Energy Outlook* (Washington, D.C.: Government Printing Office, 2002).

uses of coal have changed dramatically. In 1950, most coal was consumed in factories, homes, and trains. In 2000, half of U.S. electricity was produced by coal-fired units. However, since 1950, the average price of coal has fallen 47 percent. Indeed, coal is the least expensive of the major fossil fuels in the United States and around the world. The future of coal in America's energy sector will depend, to a large extent, on the continuing efforts to create "clean" coal. Substantial funding has been allocated to support technological improvements to make coal a more environmentally friendly fuel.

Natural Gas

Unlike coal, natural gas is relatively clean. When burned, gas produces only one third of the carbon dioxide for each unit of energy that coal produces and two thirds as much as oil. Not surprisingly, natural gas has become the unqualified favorite fuel around the world. Indeed, natural gas has recently become the fuel of choice for nearly all new electric power plants in the United States. Under existing policy, natural gas generating capacity is expected to constitute about 90 percent of the projected increase in electricity generation between 2000 and 2020.[4] This should be seen as part of a significant change in gas consumption. For decades, the industrial sector of the economy has been the heaviest user of natural gas. In 2000, industrial entities accounted for nearly half of all natural gas consumption, followed by the residential sector, which used 20 percent.[5] In recent years, very small amounts of natural gas have been reported for use in vehicles.

This skyrocketing demand raises two questions: Will the current production levels and proven reserves be able to meet growing demand? And if not, how will this increasingly widening gap between domestic production and consumption be filled? The United States holds 3.2 percent of the proven world reserves, the sixth largest in the world.[6] From the early 1980s, production of natural gas exceeded reserve additions, and the nation's proven reserves were declining. The downward trend was reversed in 1994. The relatively high levels of annual reserve additions reflect increased exploratory and developmental drilling as a result of higher prices and expected strong growth in demand, as well as productivity gains from technological improvements.[7] These reserve additions, however, are not projected to provide enough supplies to meet the growing demand. The United States was essentially self-sufficient in natural gas until the late 1980s, when consumption began to significantly outpace production. The import share of consumption rose from 5 percent in 1987 to 15 percent in 2000.[8]

Natural gas is seen as a clean and reliable resource that is expected to fuel an increasing share of U.S. energy consumption. This rosy prospect, however, faces three important challenges: location of new discoveries, the need to import increasing supplies, and the need to upgrade and establish modern pipelines to distribute natural gas.

First, most of the domestic production from traditional sites is declining. Increasingly, the nation will have to rely on natural gas from unconventional resources, such as tight sands, deep formations, deep water, and gas hydrates. The full utilization of resources in such locations requires substantial investments and sophisticated technology over several years. Second, net natural gas imports are expected to grow for the foreseeable future. Most of the rise in demand will be increasingly met by supplies from Canada, which has a very large gas resource base and easy pipeline access to the lower forty-eight states. Furthermore, liquefied natural gas is projected to provide a growing percentage of natural gas imports.[9] Finally, natural gas distribution is hindered by an aging and inadequate network of pipelines. Matching supply and demand will require the construction of thousands of miles of new gas pipelines, costing $1.5 trillion.[10]

Oil

The United States is the world's largest oil consumer and importer and the second largest producer (after Saudi Arabia). It ranks eleventh worldwide in proven reserves.[11] Oil is the nation's largest source of primary energy, providing 38 percent of energy needs. Oil is projected to maintain its share of U.S. energy consumption in spite of declining reserves. Our growing dependence on imported oil can be explained by declining production and rising consumption since 1970. The figures in table 1.2 illustrate these two trends.

Table 1.2. U.S. oil production and consumption, 1970–2000 (thousand barrels per day)

Year	Production	Consumption
1970	9,635	14,350
1971	9,465	14,845
1972	9,440	15,990
1973	9,210	16,870
1974	8,795	16,150
1975	8,375	15,875
1976	8,130	16,980
1977	9,865	17,925
1978	10,275	18,255
1979	10,135	17,910
1980	10,170	16,460
1981	10,180	15,550
1982	10,200	14,765
1983	10,245	14,745
1984	10,505	15,175
1985	10,545	15,170
1986	10,230	15,665
1987	9,945	16,025
1988	9,765	16,630
1989	9,160	16,665
1990	8,915	16,305
1991	9,075	16,000
1992	8,870	16,260
1993	8,585	16,470
1994	8,390	16,950
1995	8,320	16,950
1996	8,295	17,470
1997	8,270	17,770
1998	8,010	18,030
1999	7,730	18,635
2000	7,745	18,745

Source: BP/Amoco, *BP Statistical Review of World Energy* (London, 2001).

After oil production peaked in the United States in 1970, it started a slow but steady decline. One reason for this slow shortfall in domestic production was the surge in Alaskan North Slope oil production beginning in the late 1970s. Production from Alaska, however, peaked in the late 1980s and gradually fell to a modest level by the late 1990s and early 2000s. For the foreseeable future, U.S. oil production is projected to continue falling. Advances in exploration and production technologies do not offset declining oil resources.

On the other side, a close examination of the trend in U.S. oil consumption suggests a clear sensitivity to prices and economic performance. Following the 1973 price hikes, initiated by the Organization of Petroleum Exporting Countries (OPEC), U.S. consumption of petroleum declined for two years. Then it started to climb. The jump in oil prices in the late 1970s and early 1980s, caused by the Iranian revolution and the outbreak of the Iran-Iraq war, resulted in a marked decline in consumption. Since the mid-1980s, two factors have contributed to a steady increase in U.S. petroleum consumption: stable prices at a relatively low level and substantial economic growth in the United States.

This rise in oil consumption has been driven most notably by the transportation sector. Generally, energy is consumed in three broad end-use sectors: the residential and commercial sector, the industrial sector, and the transportation sector. While other sectors have shown some ability to substitute other energy sources for oil, transportation has not. Indeed, trucks and cars account for approximately two-thirds of U.S. petroleum use. The slow change in the automobile's use of gasoline over the past three decades reflects the sensitivity to both price jumps and strong economic performance. The average efficiency of the vehicle fleet, as measured by miles per gallon (mpg), improved from 13.4 mpg in 1973 to 21.2 mpg in 1991.[12] This improved efficiency was partly in response to the enforcement of the Automobile Efficiency Standards. In recent years, relatively low fuel prices and higher personal income have resulted in consumer demand for larger and more powerful vehicles. Thus, the growing share of small trucks and sport utility vehicles in the transportation sector began to slowly pull the mpg back to the low level. Future fleet fuel efficiency improvements will depend upon the rate of replacement of older, less efficient vehicles along with trends in consumer preferences. Finally, it is important to point out that technological advances such as gasoline fuel cells, direct fuel injection, and electric hybrids for both gasoline and diesel engines are not likely to reduce petroleum consumption. The high rate of U.S. per capita motorization (car ownership) will keep gasoline use at a high level. In short, transportation sector will continue its upward pressure on U.S. energy consumption, particularly petroleum.

This projected fall in oil production in conjunction with an expected rise in consumption will result in increased dependence on imported oil. Until the 1950s, the United States produced nearly all the petroleum it needed. But by 1960, the gap between production and consumption had begun to widen, and imported petroleum became a major component of U.S. energy needs. During the 1960s, consumption slightly outpaced domestic production, and a more significant gap had developed by the early 1970s. Between 1973 and 2000, U.S. dependence on foreign oil rose from about 35 percent to more than 52 percent of domestic consumption. This has significant negative implications for the nation's economic performance. In 2000, the U.S. energy trade deficit was about $120 billion, most of which was spent on oil imports. Without a drastic change in current policy, the share of U.S. oil demand met by imports is projected to increase to 64 percent in 2020. In other words, by 2020 the oil for nearly two of every three gallons of U.S. gasoline and heating oil could come from abroad.[13]

To sum up, in spite of efforts to diversify the United States' mix of energy, the nation still is deeply dependent on oil as the major source of energy. The share of natural gas is projected to expand, but petroleum will continue to play a major role in meeting the growing American demand for energy, most notably in the transportation sector. Given the lack of domestic oil resources, the United States will remain dependent on foreign supplies. This dependence has raised concerns regarding the nation's energy security.

The question of how to achieve a state of "energy independence" or reduce America's dependence on foreign supplies has been raised by policymakers for a long time. Before addressing this question, two important caveats should be taken into consideration. First, a distinction should be made between dependence and vulnerability. Oil dependence does not necessarily mean that the United States is vulnerable to an oil disruption. If the world oil supplies come from many producers and one of them suddenly stopped exporting oil, this would have little impact on U.S. and world economies, even at a high rate of U.S. dependence. Concentration on few suppliers, not dependence, would lead to vulnerability. Second, the globalization of oil markets has made the question of who sells and buys a particular barrel of oil less and less relevant. U.S. oil security depends on sufficient supplies to support both American and global economies. As long as Washington participates in the global oil market, it will feel the impact of supply disruptions wherever they occur. Interdependence, rather than independence, is the cornerstone of contemporary oil security.

America's Quest for Curbing Dependence on Oil

Since the 1930s, policymakers in Washington have considered articulating and implementing a national energy policy. Under Franklin D. Roosevelt's administration, there was a strong belief that the government could not solve the economic problems facing the country without playing a role in oil policy, which was considered a vital factor in the economic recovery. The intention was not to nationalize or make the industry public but to coordinate its activities. With U.S. involvement in the Second World War, the struggle over the formulation of a governmental oil policy intensified. Despite the heavy drain on its oil supplies during the war, the United States still occupied a strong position with respect to petroleum. In 1950, the United States provided 52 percent of the world's crude oil production. By 1998, that figure had dropped to about 10 percent.[14] Foreign oil has been imported into the United States in ever-increasing amounts, and the notion of "oil independence" was gradually accepted by many policymakers in Washington.

Before reviewing the main efforts to achieve this goal—oil independence or reducing U.S. dependence on imported oil—it is important to highlight four of the main characteristics of America's oil policy since the early 1950s.

First, generally speaking, U.S. oil policy can be described as "guided laissez-faire."[15] Oil companies make their own production and investment decisions based on commercial consideration. At the same time, the U.S. government has occasionally intervened diplomatically, economically, and militarily to guarantee these companies access to oil. In other words, there is some kind of coordination between the U.S. government strategic interests and the commercial interests of American oil companies. In most cases, the two sides supported each other, but in few but significant cases, as will be explained later, they went separate ways. Second, the United States has never articulated and implemented a long-term and comprehensive energy strategy. Major energy initiatives were taken, largely, to address specific crises and did not last. Third, by 1972, Americans had become accustomed to expanding energy consumption with minimal concerns about the constancy of supply or sharp price escalation. A turning point came in the early 1970s when U.S. oil production peaked and the country became seriously dependent and vulnerable to foreign oil supplies. The 1973 prices spikes and Arab boycott (to be discussed later) aggravated this sense of vulnerability. A major component of "energy security" has been to reduce the nation's dependence on imported oil and boost domestic production. The independent American producers have led the way in resisting the "invading oil" based on national security considerations. More than 8,000 independent crude oil and natural gas explorer/producers in the United States founded the Independent Petroleum

Association of America, which is dedicated to ensuring a strong and viable domestic oil and natural gas industry. Since the early 1950s, they have strongly warned Washington of the dangers of oil dependence.[16]

These independent American producers found receptive ears in Dwight D. Eisenhower's administration. Eisenhower was convinced that the growing share of imported oil in U.S. energy consumption represented a challenge to the country's national security and its prominent role in world affairs. Eisenhower's energy policy had two objectives. It aimed at reducing the share of foreign oil to roughly 12 percent of total consumption and relying more on oil supplies from Canada and Mexico than from faraway producers. Thus, after two years of requesting voluntary import quotas, which oil companies did not comply with, the president made it mandatory in 1959. Under this program the exporting countries were divided into separate groups, depending on preferential treatment. Western Hemisphere exporters were favored.

The impact of this mandatory import quotas program on U.S. oil policy was mixed. The United States became relatively independent of foreign oil reserves during most of the 1960s, which was seen as a positive development in the country's national security. Meanwhile, exports from Canada to the United States increased sharply and replaced the previously imported oil from the Middle East despite the fact that the consumers had to pay a higher price for Canadian oil. At the same time, most of the cheap American oil reserves were utilized and thereby exhausted. The program stimulated production levels that eroded domestic reserves rather than creating stockpiles and spare capacity. In the late 1960s and early 1970s, oil companies found that it was more profitable to pay additional import fees than to use domestic oil, since domestic production costs were higher than the total cost of imported oil plus the import fee.[17]

The Nixon and Carter administrations had to deal with some of the most serious oil crises. In the early 1970s, American domestic oil production began its steady decline. Consequently, the country's dependence on imported oil increased. Under these unfavorable circumstances, the first price shock and Arab oil embargo took place. The United States was ill prepared. These two related events underscored America's sense of vulnerability to disruption of foreign supplies. In a symbolic move, Richard Nixon announced that, because of the energy crisis, the lights on the national Christmas tree would not be turned on. In addition, he signed the Emergency Highway Conservation Act, setting a speed limit of fifty-five miles per hour.[18] Most important, Nixon announced a plan called "Project Independence," the aim of which was to develop domestic resources to meet the nation's energy needs without depending on foreign suppliers. He wanted to achieve a state of self-sufficiency within a decade. Project Independence proposed measures to stimulate investments in domestic oil production, including de-control of domestic energy prices and subsidizing domestic oil by imposing fees on imported oil. This unrealistic goal of self-sufficiency was never achieved.

Nixon's successor, Gerald Ford, recommended a comprehensive energy program that featured higher taxes on imported oil and the gradual phasing out of price controls that the government had placed on domestic oil.[19] Ford also came out in favor of a windfall profits tax on domestic petroleum, the decentralization of oil prices, the stockpiling of 1 billion barrels of petroleum, and an increased reliance on coal, electricity, and nuclear power. Finally, Ford signed the Energy Policy and Conservation Act, which authorized the establishment of the Strategic Petroleum Reserve.

Coming to office in January 1977, Jimmy Carter judged the energy crisis to be a national emergency and offered a program to deal with it—a program that he asked the nation to accept as the "moral equivalent of war."[20] Probably more than his predecessors, Carter focused more on the demand side than the supply side of the energy equation. His program called for reduced overall energy consumption, significantly reduced imports, increased reliance on coal and renewable sources of energy like sunlight, wind, and wood, higher gasoline taxes, and various tax credits and incentives to encourage more efficient automobiles and home insulation. Also, at the president's request, Congress created a new cabinet post, Department of Energy, in 1977.

During most of the 1970s the official objective of U.S. energy policy was to reduce dependence on imported oil. The collapse of oil prices that followed the global oil glut in the mid-1980s undermined the sense of urgency to take drastic action to control and restrain the American appetite for more energy. Throughout the 1980s and 1990s, the centerpiece of U.S. energy policy was to foster at home and abroad deregulated markets that efficiently allocated capital, provided a maximum of consumer choice, and promoted low prices through competition. Domestically, several options to address the country's growing needs for energy have been pursued. These include an increase in domestic production by relying more on technological advances, creating a national strategic stockpile, developing a nonfossil source of energy (nuclear power), conservation, and cooperating with Canada and Mexico (Hemispheric energy policy). These five strategies are not mutually exclusive; rather, they complement each other. Their common goal is to curb U.S. dependence on imported oil. The pros and cons of each strategy need to be examined.

Technological Advances

A basic component of George W. Bush's energy policy is to adopt an aggressive approach by exploring and drilling for oil in both onshore and offshore areas within the United States. This is in contrast to the Clinton administration, which was more reluctant to open public land for oil operations. The federal government owns about a third of the nation's land. These federal lands and offshore areas are believed by some industry specialists to contain most of the nation's undiscovered domestic energy resources.

The clearest contradiction between the Clinton administration and the Bush administration is focused on the opening of the Arctic National Wildlife Refuge (ANWR). Bill Clinton strongly urged Congress to resist allowing oil drilling in "one of the last truly wild places on earth." The area is located between the Prudhoe Bay of the Alaska North Slope and the Mackenzie River Delta of Canada along the Beaufort Sea. Some analysts believe that the ANWR contains the largest onshore, unexplored, potentially productive geologic basins in the United States. Oil production in Alaska has entered a period of decline, which can be reversed by opening up the ANWR. Proponents of President Bush's energy plan claim that such an opening could lead to the development of resources that could make a significant contribution to domestic supply and would also bolster domestic industry and the local and national economies. Furthermore, they argue, advances in exploration and drilling technology would cause very little, if any, harm to the region's delicate ecosystem. In response to a congressional request early in 2000, the Department of Energy issued a report on potential oil reserves and production from the ANWR. The report projected that there are approximately 10.3 billion barrels of oil technically recoverable. The report added that ANWR's peak production rates could range from 1.0 to 1.35 million barrels per day (b/d), with initial ANWR production possibly beginning around 2010 and peak production twenty to thirty years after that.[21]

In addition to ANWR, production from offshore sources, particularly the Gulf of Mexico, is predicted to play an increasingly important role. The first offshore well was drilled in 1938, one mile from the Louisiana coastline. Recent technological advances have made it possible to economically produce oil from nonconventional sites. Examples of these advanced technologies in oil exploration include three-dimensional seismic technology that enables geologists to use computer images to determine the location of oil and gas before drilling begins, dramatically improving the exploration success rate; deep-water drilling technology that enables exploration and production of oil and gas at depths over two miles beneath the ocean's surface, and highly sophisticated directional drilling that enables wells to be drilled long horizontal distances from the drilling site. These technologies have drastically changed the practices and capabilities of the oil industry, particularly in exploration costs and success rates. A new barrel of oil reserves that cost about $15 to find in the 1970s now costs less than $5.[22] Similarly, drillers are more successful in locating oil wells than they were in the 1970s.

In spite of these significant potentials to increase domestic production by relying on modern and highly sophisticated technological advances, three restraints should be taken into consideration. First, the unpredictable ups and downs of world oil prices would play an important role in the speed and magnitude of oil exploration in ANWR, the Gulf of Mexico, and other geologically complex reservoirs. In spite of substantial cost reduction, it is still cheaper to produce in foreign countries (particularly the Persian Gulf) than in the United States. Thus, high oil prices will make it cost-effective to explore for oil in the United States, and low prices are likely to slow down domestic production. Second, despite drastic changes in the American oil industry, technological advances have not led to any significant change in the production decline that has been under way since the early 1970s. Technology cannot make up for declining resources. Finally, most of these untapped oil reservoirs are located in environmentally sensitive areas. The strong opposition to drilling by environment activists should not be underestimated. This opposition might completely prevent or slow down oil exploration, particularly in ANWR. The Bush administration's assertion that sophisticated technology can guarantee energy production without harming the environment has not been accepted by all participants in the energy debate.

Conservation

Conservation implies less useable energy input per output of useful energy and has been a constant theme of interest in the political economy of energy. In the 1970s, a world fearing energy shortage prompted concern over investment in energy conservation as an alternative to investment in supply. More recently, interest in conservation was revived by concern over environmental damage from energy use (e.g., global warming).[23]

The U.S. economy has become more efficient in its overall use of energy. Energy efficiency is the ability to use less energy to produce the same amount of lighting, heating, transportation, and other energy services. One measure of energy efficiency is energy intensity—the amount of energy it takes to produce a dollar of gross domestic product. Since the early 1970s, the U.S. economy has grown by 126 percent while energy use has increased by 26 percent.[24] In other words, the American economy has grown nearly five times faster than energy use. (Europe and Japan made even greater gains in efficiency.) Three factors contributed to this improvement. First, the price spikes of the 1970s forced many people to use less oil or shift to other forms of energy (natural gas and nuclear power). This was particularly evident in the industrial and residential sectors and to a lesser degree in the transportation sector. Second, the growing concern in the United States and around the world in reducing pollution and protecting the environment has provided another incentive to cut oil consumption. Third, there has been a fundamental change in the structure of the American economy. The shift from manufacturing to services has contributed significantly to reduced energy consumption and improved efficiency.

Since the mid-1980s, the decline in energy intensity has moderated. The reason for this less-efficient use of energy is twofold. The strong economic performance has led to expansion in consumption. This includes larger homes, bigger cars, and more air travel. On the other hand, the stability of oil prices at a relatively low level has encouraged more fuel consumption. Thus, between 1970 and 1986, energy intensity declined at an average annual rate of 2.3 percent, whereas it declined by only 1.3 percent between 1986 and 2000.[25] In spite of this improvement, it is important to emphasize that conservation alone is not the answer. Rather, conservation should be integrated in a comprehensive and long-term energy strategy. The increasing reliance on a nonfossil energy source—nuclear power—is another component of this strategy.

Nuclear Power

George W. Bush's energy plan calls for a fresh look at nuclear power as a significant contributor to the long-term national sustainable energy mix. Nuclear energy is used exclusively to generate electricity, providing the second largest source of U.S. electrical energy.[26] Environmental concerns are the reasons for relying on nuclear power. Unlike fossil fuels, nuclear power produces no emissions (nitrogen oxides, sulfur dioxide, mercury, and carbon dioxide). Given these features as well as the growing awareness of U.S. dependence on imported oil, there was a strong enthusiasm for nuclear power in the 1960s. The first commercial nuclear energy–powered facility went into operation at a site on the Ohio River in Shippingport, Pennsylvania, in 1957. Many nuclear plants were constructed, licensed, and connected to the electricity grid. Most of these nuclear power plants received forty-year operating licenses and are scheduled to retire around 2015. However, this wave of enthusiasm for nuclear power was broken by the accident at Three Mile Island in 1979, which reinforced safety concerns.[27] Since then, there have been no new reactor orders, and none are currently planned. As operable nuclear power plants have aged, some have become too expensive to operate and have been closed. The joint effect of shutdowns and lack of new units is that the number of operable nuclear plants has fallen drastically.

Despite these grim prospects for the future of nuclear energy, it is important to point out the improved performance of nuclear energy plants since the early 1980s. These plants generate more electricity at a lower cost. Furthermore, their safety record has improved. The Nuclear Regulatory Commission rigorously oversees the operation and maintenance of these plants. In 2000, it granted its first renewal-operating license to a nuclear power plant.[28]

Due to these favorable conditions—reduced costs and improved safety records—the decline in nuclear energy generation has not been as rapid as was predicted. Nuclear energy is likely to continue providing a share of U.S. energy needs. Still, a significant challenge to the future prospects of nuclear power is how the industry would handle the questions of safety and radioactive waste. While the U.S. nuclear industry has seen years of seemingly trouble-free operation, nuclear power's global experience does not mirror this trend. More public confidence needs to be gained. Furthermore, a solution to the problem of radioactive waste has yet to be found. In the early 1980s, most of the U.S. nuclear reactor operators signed a contract with the federal government to take the fuel to a proposed permanent waste facility at a deep geological site in the remote desert at Yucca Mountain in Nevada.[29] However, environmental concerns, including questions about possible earthquakes at the site and the safety of routinely hauling the material cross-country, have prevented the final implementation of this agreement.

The Strategic Petroleum Reserve

The best insurance against interruptions in petroleum supplies is having a large stock of replacement that the government can release promptly. Thus the Strategic Petroleum Reserve (SPR) was created and is seen as the nation's first line of defense in case of an oil crisis. The need for a national oil storage reserve has been recognized for at least five decades. Secretary of the Interior Harold Ickes advocated the stockpiling of emergency crude oil in 1944. President Truman's Minerals Policy Commission proposed a strategic oil supply in 1952. President Eisenhower suggested an oil reserve after the 1956 Suez crisis. The Cabinet Task Force on Oil Import Control recommended a similar reserve in 1970.[30] These proposals were finally implemented in the aftermath of the 1973–74 oil embargo. The price shock and the embargo aggravated America's sense of vulnerability and created the right conditions to move ahead with the plan to establish a national oil storage. President Ford set the SPR into motion when he signed the Energy Policy and Conservation Act (EPCA) in 1975. The legislation declared it to be U.S. policy to establish a reserve of up to 1 billion barrels of petroleum. The Gulf of Mexico was chosen for oil storage sites because it is the location of many U.S. refineries and distribution points for tankers, barges, and pipelines.

The SPR proved its value in 1991 when President George Bush ordered the first-ever emergency drawdown of the SPR. This step contributed to the stability of the world oil markets and prices. Two more public sales of crude oil from the SPR were held in 1996 and 1997. In 2000, President Clinton authorized another public sale in order to bolster the U.S. oil supplies and to alleviate possible shortages of heating oil during the upcoming winter. Finally, in the aftermath of the September 11 terrorist attacks against the United States, George W. Bush ordered the Department of Energy to fill the SPR to its capacity (700 million barrels) over the next few years. This will be done through the so-called in-kind payments. In other words, instead of taking money from the companies that lease federal lands for oil and gas drilling, the government will take oil. Given the high degree of uncertainty in the international system (e.g., the war on terrorism), this unprecedented initiative should strengthen the United States' insurance against a break in the flow of oil shipments.

Despite the fact that the SPR is considered the federal government's major tool for responding to oil supply disruptions, two problems can be identified. First, until President Bush decided to fill the SPR to its capacity, the Reserve had declined both as a share of imports and in absolute size. The volume of oil stored in the SPR peaked at 592 million barrels in 1994. By 2001, the SPR contained around 541 million barrels of oil.[31] This decline is more dangerous when measured by the number of days of net oil import protection. The SPR shrank from 115 days of import replacement in 1985 to around 54 days in 2001.[32] Second, under the EPCA there was no preset "trigger" for withdrawing oil from the SPR. In 1991, the withdrawal was mainly in response to potential interruption of supplies due to the Iraqi invasion of Kuwait, while the sale in 2000 was initiated to dampen price hikes. Since its inception, the SPR has been used by policymakers both as a tool of crisis management and as an instrument to counter high oil prices. Policymakers seeking price mitigation walk a fine line between "calming" the market by showing that there is sufficient crude available and yielding the unintended consequence of short-circuiting the price mechanism and preventing the market from equilibrating.[33] This dilemma is magnified in political debate, pitting the advocates of free markets against the advocates of interventionist government. A clear policy for the use of the SPR should be established.

Hemispheric Energy Policy

Establishing strong energy ties between the United States and its neighbors to the north and the south has been considered by several administrations as a cornerstone in ensuring American's energy security. President George W. Bush has made creating and strengthening a hemispheric energy policy a priority for addressing the nation's deepening dependence on imported hydrocarbon resources. Indeed, Canada and Mexico have been major suppliers of oil to the United States for several years. (Other major suppliers have included Saudi Arabia, Venezuela, and Nigeria.)

The interdependence between the three North American states is even stronger in natural gas than in oil. Given the high costs of transporting gas, the fuel has always been labeled as a "continental commodity" and is traded mostly within the same region where it is produced rather than cross-continents as oil, which is cheap and easy to transport. Thus Canada, the United States, and Mexico have intensified their efforts to create an integrated natural gas market. Cross-border pipeline capacity between the three countries is increasing, export/import activities are growing, and prices are converging. The driving force behind the growth in all three countries is the increased consumption of natural gas for electric power generation. The American-Canadian energy relations are different from those with Mexico due to variation in resource base and fiscal policies in each country.

The past few years have witnessed a substantial expansion of pipelines between the United States and Canada. Particularly significant are the Northern Border system through Montana into the Midwest, the Alliance Pipeline through North Dakota into Chicago, and the Maritimes and Northeast system to New England markets.[34] Not surprisingly, Canada supplies most U.S. natural gas imports and is expected to continue. Canada's vast hydrocarbon resources and relatively easy access to the American market explain this extensive energy trade. The construction of new pipelines might slow down due to environmental sensitivities.[35]

Trade between the United States and Mexico has substantially expanded in the past several years. While U.S. oil and natural gas exports to Canada are negligible, the United States is a net exporter of natural gas and refined petroleum products to Mexico. Two main characters of the Mexican energy industry, particularly the natural gas sector, pose a challenge to close cooperation with Washington. First, Mexico's domestic consumption of natural gas is rapidly increasing, and its indigenous production is not keeping pace with the demand. For the foreseeable future, the country is likely to remain a net importer. Second, Mexico's energy industry is run by a state monopoly, Petroleos Mexicanos. In 1938 Mexico became the second country to nationalize its oil industry. (The first was the Soviet Union, in 1917.) Mexican laws prohibit foreign or private ownership in the energy sector. The country needs to invest billions of dollars to upgrade and modernize its oil and gas infrastructure. Petroleos Mexicanos simply does not have this money, and there is no substitute for opening the door to private and foreign investments. In 1995, President Ernesto Zedillo gave a new policy orientation to govern the natural gas industry: Private investment was to be encouraged in gas transportation, storage, and distribution. Five years later, very little has changed.[36] President Vicente Fox, who took office in December 2000, indicated his support for economic reform. But first he must persuade Mexico's Congress to rewrite the laws to allow a mix of public ownership and private and foreign participation in energy production.

Finally, it is worth emphasizing that turning to Hemispheric sources of oil to reduce U.S. vulnerability to oil crises should not be seen as a guarantee for energy security. America's main trade partners in Europe and Asia still depend heavily on oil supplies from other producers, particularly from the Middle East. Interruptions of supplies to these important trade partners will not only affect their economies but will negatively impact the American economy.

Conclusions

Political instability in the Middle East has convinced the United States and other major oil-consuming countries to search for oil resources in other regions. Thus, since the mid-1970s, exploration and development operations have accelerated all over the world, including the North Sea and the Caspian Basin. Recently, South Atlantic Region, which combines both the Atlantic Equatorial Africa and South America, has emerged as a new oil supply flashpoint. African oil tends to be of high quality and low in sulfur, making it suitable for stringent refined product requirements and giving it a growing market share for refining centers along the east coast of the United States. Similarly, Latin American and Caribbean energy producers such as Venezuela, Brazil, and Trinidad and Tobago have expanded their share of the U.S. oil and natural gas market.

Despite the increasing supplies from the North Sea, Latin America, southern and western Africa, and the projected contribution of the Caspian Basin to global oil security, the Persian Gulf remains the main area that can meet either a substantive increase in demand or an emergency caused by a major disruption of supplies. Specialists in the oil energy industry use a measure called reserve to production, which is the time reserves are expected to last at current production rates. In 2000 the reserve to production ratio in Europe was 8.3 years, North America 13.8, Asia Pacific 16.3, former Soviet Union 24.2, Africa 28.2, South and Latin America 37.7, and the Middle East 87 years.[37] In other words, the Gulf region will continue to occupy the driver's seat in the oil industry. This can be explained by at least four facts. First, the Gulf states hold more hydrocarbon resources than any other producers. In 2000, the Persian Gulf region contained around 672 billion barrels of proven oil reserves, representing approximately 65 percent of total world oil reserves and 1.8 trillion cubic feet of natural gas reserves (34 percent of the world total). Second, in 2000 the Gulf region maintained a significant percentage (44 percent, or 1.6 million b/d) of the world's excess oil production capacity.[38] This is important because in the event of an oil supply disruption, excess oil production capacity can be brought on-line to compensate for the lost supplies. If such a disruption were to occur in the Persian Gulf, it would leave the world with relatively limited options for making up the lost oil production. Third, oil can be extracted from the ground in the Gulf at a lower cost than in any other region. The marginal costs of production in the Gulf states are usually just a fraction of current prices, which means that it is more profitable to produce oil in the Gulf region than anywhere else. Fourth, oil fields in the Gulf are located close to the international markets and enjoy well-developed transportation routes. Most oil shipments from the region are not restrained by any financial or political barriers. Given all these advantages, all indicators point to increasing reliance by oil-importing countries on Gulf producers. Indeed, dependence on oil

Table 1.3. Imports from the Persian Gulf region as a percentage of net oil imports, 1983–2000

Year	United States	Western Europe	Japan
1983	8.8	41	60
1984	9.3	39	61
1985	6.1	35	59
1986	14.7	45	58
1987	16.1	43	60
1988	20.8	44	58
1989	23.1	47	63
1990	24.5	48	65
1991	24.2	43	64
1992	22.5	43	66
1993	20.7	50	68
1994	19.2	48	68
1995	17.8	47	70
1996	16.9	43	70
1997	17.3	47	74
1998	19.9	50	76
1999	23.0	50	74
2000	22.1	45	75

Source: Energy Information Administration, *Persian Gulf Oil and Gas Exports Fact Sheet* (February 2001), on-line at www.eia.doe.gov.

supplies from the Persian Gulf region has deepened over the past two decades in the United States and other industrialized countries, as table 1.3 illustrates.

The Persian Gulf's share in the net oil imports of industrialized countries has risen since the mid-1980s. This can be explained by the sharp decline of oil prices early in the decade. In other words, after almost ten years of volatile and high oil prices, the oil glut of the 1980s, which stabilized prices at a relatively low level, has contributed to deepening dependence of American and world economies on oil supplies from the Persian Gulf. For the foreseeable future, this trend is not likely to be reversed. Indeed, all indicators point to the Persian Gulf's expanding role in the global oil market.

To meet this growing demand for oil, production capacity has to be expanded in major Gulf producers. Many of these producers have considered opening their oil industry to foreign investment. This important development may have significant implications for U.S. and world oil security. It also shows that expanding production in the Persian Gulf states will not be restrained by geology. The region is rich in hydrocarbon resources.

Rather, political factors will play a crucial role in determining the scope of this necessary expanding production capacity. Critics of American government claim that Washington has made energy goals secondary to other foreign policy objectives, particularly during the 1990s. American sanctions policy has slowed development of plentiful resources in Iran (and Libya), while Iraqi production has been held back by the United Nations sanctions regime. The role of sanctions in constraining investment in several key Middle East producers has meant less diversification of sources and less total capacity. Meanwhile, domestic political and economic considerations in Saudi Arabia have added a sense of uncertainty to the opening of the kingdom's oil industry to foreign investment.

Thus, by any estimation, Persian Gulf oil producers will remain central to world oil security, and the region will continue to be a primary focus of U.S. energy policy. The September 11 terrorist attacks affected almost all aspects of life in the United States, including the nation's energy security. Within a few hours after the attacks, security at U.S. nuclear power plants increased dramatically. Also, the demand for jet fuel fell sharply, and the concern over the possible interruption of oil supplies prompted the Bush administration to take steps to fill the SPR to its full capacity. Equally important, the terrorist attacks changed some of the parameters of American foreign policy toward both the Persian Gulf and Central Asia regions based on their cooperation, or lack of it, in the war on terrorism.

The next chapter addresses the broad context of global energy security. Specifically, the analysis defines the major players in the global energy market. These include the European Union, Russia, and East Asia (particularly China and Japan). Finally, the chapter discusses the international community's efforts to respond to climate change. The discussion underscores the need to adopt a collective approach to energy security. The following three chapters examine energy relations between Washington on one side and Riyadh, Tehran, and Baghdad on the other side. In addition, American energy policy in the Caspian Sea states will be analyzed. Given these two regions' current and projected hydrocarbon wealth, they represent absolute strategic importance to the United States.

In closing, several of the major findings of this study need to be highlighted. First, for a long time the United States has not had a long-term and comprehensive energy policy. The solution to what many analysts perceive as "energy crisis" will take some time. In addressing the energy question, persistence, not urgency, is needed. Second, an integrated and balanced approach should be adopted to the energy problems. This approach should take into consideration both the supply and demand sides. This approach should also be built on diversity of both the fuel mix and the geographic origin of that fuel. Third, in spite of this need for diversification, fossil fuels, particularly oil and natural gas, will continue to dominate U.S. energy needs. Fourth, oil security of any country does not mean achieving a state of self-sufficiency. Given the declining production, limited reserves, and rising demand, the United States will continue to be dependent on oil supplies from abroad. The goal, therefore, is not to engage in a meaningless debate over energy independence but to find new ways of managing dependence. Fifth, the heavy interdependence between the American economy on one side and those of Europe, Japan, and other Asian nations means that U.S. national energy security depends on sufficient supplies not only to the American market but also to those of major U.S. trade partners. Accordingly, it matters little if Gulf oil flows to Asia or to the United States. The global nature of oil trade and pricing means that an oil crisis would affect all. Sixth, achieving a state of stability in oil prices and security of supplies would benefit both producers and consumers. The global energy market has shifted from the confrontational attitude between consumers and producers in the 1970s to one of cooperation in the 2000s. Oil security is seen less in zero-sum terms and more as mutual benefits between consumers and producers.

2

The Global Energy Scene

Oil remains the largest single category in international trade, whether measured by volume or value. The global energy scene is fundamentally different from what it used to be. There are more producers, consumers, and multinational oil and gas companies. Furthermore, the relative significance of many players has substantially changed. In short, the energy market is increasingly globalized, transparent, and computerized. Within this context, three characteristics of the global energy scene can be identified.

First, for a long time the question of energy security was narrowly defined as reducing dependency on any single region (e.g., the Persian Gulf). The interruptions of oil supplies from the Middle East in the 1970s and 1980s strengthened this approach. The producers' and consumers' interests were perceived as mutually exclusive. One side's gains were considered the other side's losses. In today's market environment, energy security is a shared issue for all players in the global markets. The worry among consumers over the availability of supplies is correlated with the producers' concern for secure markets for their oil and gas. In addition, multinational oil and gas companies are employing their financial assets, management expertise, and technology, side by side with producing governments, to ensure stability in the global energy markets. Finally, producers, consumers, and the companies share an important concern—containing and reducing global pollution.

Second, major oil- and gas-consuming regions have developed strategic partnerships with their close producing neighbors. These include the "Hemispheric Energy Policy" between the United States and Canada and Mexico, Western Europe and Russia, and East Asia and the Persian Gulf. This increasing "regionalization" of oil and gas movements reduces shipments expenses and paves the way for cooperation in other areas.

Third, in spite of this growing "regionalization," it is important to point out that oil is fungible and moves to buyers willing to pay the "right"

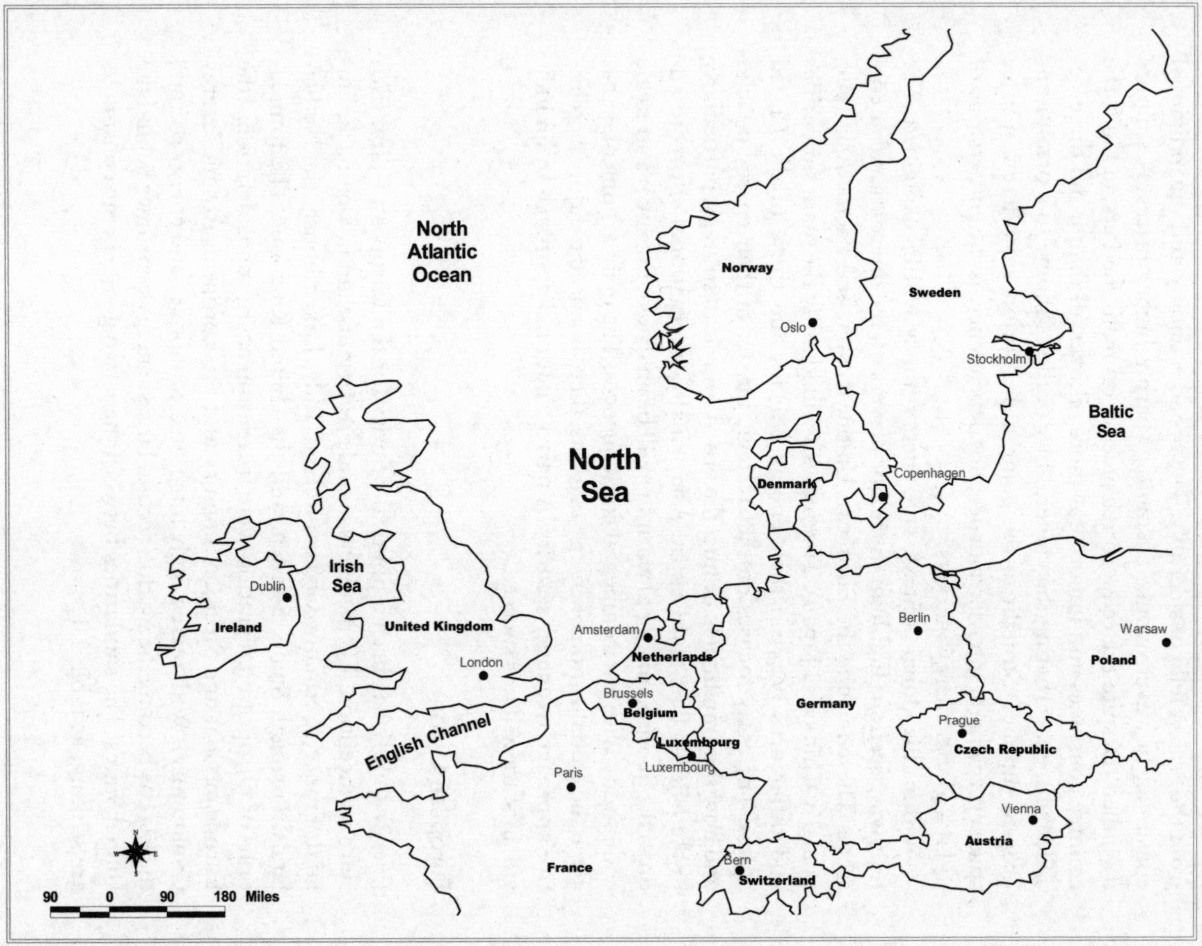

Map 2. The North Sea and north-western Europe

price. The real adverse impact of oil supply interruptions has nothing to do with the level of dependence on any specific region but on any effect that supply shortfall will have on prices. Thus, since the United States imports a smaller percentage of its oil from the Persian Gulf than it did in the early 1970s, it would not be paralyzed by oil shortages if supplies from the region fell considerably. The broad economy and the individual consumer, however, would encounter sharply higher prices as supplies dwindled. Despite the reduced reliance on oil from the Persian Gulf, the country consumes and imports far more oil over all than it did three decades ago, exacerbating the vulnerability of the economy to a precipitous drop in supplies. Straight to the point, a disruption anywhere is a price spike everywhere. Interdependence, not dependence, is the cornerstone of today's global energy markets.

Within this global context, U.S. energy policy will be examined. The following sections highlight the major players on the international energy scene. These include the European Union, Russia, and Pacific Asia (particularly China and Japan). The analysis will identify the main similarities and differences between their energy policies and those of the United States. The essay focuses on the strategic impact of their energy policies and how they might help or hurt the broad American strategy, particularly in the Middle East and Central Asia. Finally, international efforts to address the question of global warming will be analyzed. Energy is a strategic commodity with environmental consequences. The main argument is that given the globalization of energy markets, there is a need to adopt a collective approach to energy security. A narrow national approach by an individual state will not succeed.

European Union

Energy supply has been a political priority for the European Union (EU) since its inception. EU members include Austria, Belgium, Denmark, Finland, France, Germany, Greece, Ireland, Italy, Luxembourg, the Netherlands, Portugal, Spain, Sweden, and the United Kingdom. The foundations of European integration were treaties on energy considerations (the European Coal and Steel Community and the European Atomic Energy Community). With increasingly integrated economies and energy sectors, the EU has become the world's second largest energy consumer behind the United States. The similarities between these two giant energy consumers are striking, as table 2.1 shows.

Table 2.1. Fossil fuels in the United States and the European Union, 2000: Proven reserves, production, and consumption as percentage of total world

	Proven reserves			Production			Consumption		
	Oil	NG	Coal	Oil	NG	Coal	Oil	NG	Coal
U.S.	2.8	3.2	25.1	9.8	22.9	26.7	25.6	27.2	25.8
E.U.	0.7	2.2	7.4	5.0	9.0	8.0	18.0	16.0	11.0

Sources: BP/Amoco, *BP Statistical Review of World Energy* (London, June 2001); Energy Information Administration, *Regional Indicators: European Union*, on-line at www.eia.doe.gov.

The fossil fuel gap between production and consumption in the EU is larger than that in the United States, particularly when Europe's poor reserves are taken into consideration. In other words, the EU is more dependent on foreign energy supplies than the United States. A close look would shed light on how Europe has sought to deal with this growing dependence on imported oil and gas supplies.

Oil has a larger share of the energy market than any other fuel type in the European Union. This proportion, however, is falling. The interruptions in oil supplies following political upheavals in the Middle East in the 1970s and 1980s have underscored Europe's sense of vulnerability and prompted policymakers to invest in and develop other sources of energy besides oil. As a result, the use of oil for power generation has been partly phased out, but petroleum remains the dominant fuel for the transportation sector. In short, despite sincere and successful efforts to reduce dependence on oil as the main source of energy, European economies still depend heavily on crude products. Given the limited proven reserves, the share of imported oil in the total EU's oil supplies is projected to increase from 80 percent in 2000 to 90 percent by 2020.[1] This dependence would further deepen Europe's exposure to price fluctuations. To avert such vulnerability, EU members have substantially increased their reliance on natural gas.

In addition to its contribution to the diversification of European energy mix, natural gas has three significant advantages. First, due to its chemical composition, gas has lower greenhouse emissions than oil and coal for many types of energy services, so gas causes less pollution. Second, natural gas has a geographical advantage. Most of continental Europe is already linked to gas fields in the North Sea. In other words, the necessary infrastructure to distribute gas is already there. Finally, natural gas is available in huge quantities from nearby reservoirs in Russia, Norway, and the Middle East/North Africa region. This helps in reducing costs to build pipelines. Furthermore, technological advances have substantially brought costs down along the whole liquefied natural gas (LNG) chain. As a consequence, LNG supplies are increasingly becoming competitive. This means that the EU can import gas from distant producers at reduced prices.

The North Sea holds Europe's largest oil and natural gas reserves and is one of the world's key non-OPEC producing regions. North Sea oil and gas were first discovered in the 1960s, but production grew as major discoveries continued into the 1990s.[2] Exploration and development operations in the North Sea reflect the unique character of the reservoir. On one side, production relies on sophisticated offshore technology. Consequently, the region is a relatively high cost producer. On the other hand, the region's political stability and its geographical proximity to major European consumer markets have allowed it to play a major role on the global energy scene. Accordingly, half of the increased non-OPEC oil supply during the 1990s came from the United Kingdom and Norwegian parts of the North Sea.[3] Despite these advantages, the North Sea is increasingly incapable of satisfying European oil and gas needs. Sophisticated technology has extended the possibilities for extracting oil from older or small wells or from difficult sites, but the whole reservoir is considered to be increasingly "mature," with few large discoveries likely to be made. Thus, Europe is growing more dependent on foreign supplies.

Norway is now the biggest oil exporter to the EU, while Russia, Norway, and Algeria are the main gas suppliers. The Russian share in the European market will substantially rise when the enlargement plan is implemented. Since the mid-1990s, the EU has been involved in negotiations to determine the terms of adding new members. These candidates are Turkey, Cyprus, Malta, and ten East European and former Soviet Republics: Bulgaria, Czech Republic, Estonia, Hungary, Latvia, Lithuania, Poland, Romania, Slovakia, and Slovenia. Russia dominates the energy market of these East European and former Soviet states. In addition to this growing dependence on Russia, the Middle East's share in the European energy market is expected to rise, given the region's massive hydrocarbon reserves and its proximity to the European market.

The EU region is a net importer of energy. Naturally, dependence varies by fuel and from one country to another. Growing import is not in itself a threat to Europe's energy security, but it highlights the importance of good trade links, communication, and political relationships with external partners. This is a major consideration for policymakers in Europe.

Germany, Italy, and France are the European Union's largest net energy importers. Not surprisingly, they are leading the way in building and strengthening long-term strategic partnership with major oil- and gas-producing regions. Probably more than any other Western country, Paris has always had a "special relationship" with Baghdad. In the mid-1970s, France provided modern and sophisticated weaponry systems to Iraq. Since the 1991 Gulf war, the French position in the Security Council of the United Nations has been less hostile toward Iraq than those adopted by Great Britain and the United States. Finally, French (and other European) companies have negotiated agreements with the government in Baghdad to develop oil and gas fields once the United Nations sanctions are lifted.

Like the United States, Europe understands Iran's geopolitical and geo-economic significance. But unlike Washington, Brussels believes that the dynamics of the domestic situation in Tehran make it vital for dialogue to continue so that the direction of change may be influenced. Both the United States and the EU have expressed strong opposition to some Iranian policies (e.g., violation of human rights, opposition to the Middle East peace process, sponsoring of terrorism, and efforts to acquire and develop unconventional weapons). In order to change these policies, Washington has adopted a policy of containment, whereas Brussels has relied on engagement since the early 1990s. In line with this European strategy, President Khatemi was the first Iranian leader since the revolution to visit several European states, including Germany, France, and Italy. Furthermore, the EU is Iran's largest trading partner. Finally, since the late 1990s, several European companies have been working in Iran to develop its oil and gas reserves.

Similarly, since 1988, the EU has engaged in a dialogue with the Gulf Cooperation Council, whose members are Bahrain, Kuwait, Oman, Qatar, Saudi Arabia, and the United Arab Emirates. Unlike the case with Iraq and Iran, the United States is the main political and economic partner for most of the GCC members, particularly Saudi Arabia. Since the Gulf war, Washington has become the de facto guarantor of Gulf security. Still, Gulf leaders have shown interest in maintaining and developing close relations with European powers and pursued what some analysts call a multidependence strategy (e.g., not putting all their eggs in one basket).[4] A good illustration of this strategy is the 1985 arms contract known as the al-Yamamah deal, worth $30 billion, the largest single defense contract in

British history.[5] Finally, European companies continue to compete with their American counterparts for investment opportunities in the Gulf monarchies' oil and gas sectors.

In closing, two conclusions need to be highlighted. First, there is no one single European energy policy in the Persian Gulf. Rather, the French, British, Italian, and other European interests compete with each other as well as with the United States. Second, European efforts to develop long-term energy partnerships with key suppliers and to capture their own slices of the global market face strong and growing competition from other sources, particularly Pacific Asia and Russia.

Russia

Russia holds massive hydrocarbon reserves and is the world's largest natural gas exporter and the second largest oil exporter (after Saudi Arabia). In addition, Russia is strategically located close to the fast-growing consuming regions of Europe and East Asia. No wonder energy accounts for a substantial share of Russia's exports and its overall gross domestic product (GDP). Despite this leading role, Russia's oil and gas industry experienced a severe decline for most of the 1990s as a result of the breakup of the Soviet Union. However, the industry has recently shown significant signs of recovery. Several developments have contributed to this apparent improvement in Russia's energy sector.

(1) The steep depreciation of Russia's currency, the ruble, in the aftermath of the August 1998 financial crisis. This monetary initiative increased the competitiveness of Russia's exports.
(2) The sharp rise in oil prices in 1999 and 2000 provided badly needed extra cash to the Russian economy.
(3) The election of Vladimir Putin as president in March 2000 put an end to a long period of speculation and brought a sense of stability to the political and economic systems in Moscow.
(4) Shortly after taking office, Putin embarked on an ambitious plan to reform Russia's oil and gas industries.

In order to understand these efforts to reconstruct the Russian energy sector, three interrelated characteristics should be identified. First, the country holds vast proven reserves estimated at 48.6 billion barrels of oil, 4.6 percent of the world's total.[6] Second, aging equipment and poor methods of exploration and production are employed to develop the country's oil fields. Third, the oil industry as a whole suffers from an acute shortage of investment due to the confusing tax and legal environment.[7] Accordingly, one goal of the reform process is to create a stable and attractive investment climate.

These efforts began in 1993 when the government initiated a two-step privatization process. The first phase, which involved organizing state-owned enterprises as joint-stock companies, ended in 1994 and resulted in the creation of a group of large oil companies operating in an increasingly competitive market. The second phase, which began in 1995, involves the auctioning off of large chunks of government shares in these companies.[8] Despite some improvements, foreign investment has been minimal due to an inadequate legislative framework. The Duma, Russia's lower house of Parliament, has yet to approve the final form of a national production-sharing agreement and a tax code that would regulate investment in the country's oil sector.

Given that Russia holds the world's largest natural gas reserves and is the world's leading exporter, gas industry is a major contributor to the Russian revenues and overall economy. The gas sector is almost exclusively dominated by Gazprom. The company is Russia's largest earner of hard currency, and its tax payments account for around 25 percent of federal tax revenues.[9] Consequently, a main objective of the reform process is to break the monopoly Gazprom has over this vital sector and open the door for other producers. The Putin administration has taken two important steps to achieve this goal. First, the government ordered Gazprom to give other companies the right to use up to 15 percent of its pipeline capacity. Opening the country's pipeline network to competition gave other Russian companies incentive to develop their gas and increase exports. Second, Gazprom's board of directors replaced chief executive Rem Vyakhirev with Aleksei Miller, a Putin ally. Some analysts described this management change as the end of an era for Gazprom and a step toward breaking Gazprom's upstream operations into separate producing companies in order to foster competition in the gas sector.[10]

To sum up, a significant restructuring process of the oil and gas sectors has been in place in Russia since the early 1990s and has gained momentum since Putin took office. The objectives include making the industry more competitive, attracting foreign investment, and establishing long-term partnerships with major consuming regions, particularly in Europe and East Asia, as well as expanding and solidifying Russia's strategic and economic interests in the Persian Gulf and Central Asia.

A significant move in this direction is the creation of an energy partnership with Western Europe, the Energy Charter Treaty, signed in Lisbon in

December 1994.[11] The aim is to build on the convergence of interests between Russia and the EU in the area of energy cooperation. Russia needs a substantial injection of foreign capital and technological expertise to increase its oil and gas production. Government estimates suggest that around $17 billion should be invested in the Russian gas sector and $15 in its oil sector.[12] On the other hand, with rising energy consumption and declining indigenous production, the EU's dependence on foreign oil and gas supplies is growing. In addition, Brussels would like to diversify the sources of its energy and reduce its dependence on one or a few suppliers. Finally, Europe has the financial resources, managerial expertise, and sophisticated technology that Russia needs. The two sides have agreed that investments will have to be made by private companies. Therefore, the role for the EU and Russian governments is to create the right legal framework for foreign investment in Russia.

In order to increase Russia's ability to export oil and gas to Europe, several pipeline schemes have been proposed, negotiated, and built. For example, the Baltic Pipeline System gives Russia a direct outlet to northern European markets. It became operational in late December 2001.[13] Similarly, in October 2000, Putin announced that Gazprom and Gaz de France had concluded a deal to construct a pipeline across Belarus, Poland, and Slovakia that would ship Russian gas to consumers in Western Europe.[14] Finally, in April 2001 Gazprom signed an agreement with Finnish and German customers for a feasibility study on a pipeline, North TransGas, that would carry Russian gas across the Baltic Sea.[15]

Besides this growing partnership with Europe, Russia is consolidating its energy ties with the expanding markets in East Asia, particularly in China and Japan. In order to supply China's increasing oil demand and boost its own export potential, Moscow has negotiated with Beijing to build an oil pipeline linking the two countries. In July 2000, Putin and Chinese president Jiang Zemin signed a memorandum of understanding on a feasibility study for a potential oil pipeline between the two countries.[16] In addition, the two sides have negotiated several proposals to export Russian gas to China. However, some analysts argue that it is more likely that Beijing would first exploit its own substantial domestic gas reserves before importing Russian gas.[17]

Japan is the demand key to any Russian Far East gas project. Tokyo's current and projected demand justifies Russian export pipeline projects. Since the mid-1990s, the two sides have negotiated and signed several agreements, including Sakhalin-I and Sakhalin-II, that would transport Russian oil and gas to Japan. ExxonMobil and Royal Dutch/Shell lead the consortia in charge of these schemes.[18]

These ambitious efforts to restructure the Russian oil and gas industry both at home and abroad have long-term effects on foreign policy. Several trends can be identified. First, the Russian energy industry has been slowly and firmly integrated in the international economic system. The development of Russia's production and export depends, to a large extent, on European and American investments. This growing dependence of Russia's largest and most dynamic revenue-producing industry on Western investments and the stake in the stability and long-term globalization of its activities will have a moderating effect on the industry's foreign policy influence. This trend should prevent some openly nationalistic and overtly anti-Western policies and promote more flexibility and compromise within the political establishment in Moscow.[19]

Second, the rise of Russia's oil and gas production and exports has consolidated the country's stand on the global energy scene. Increasingly, Moscow has a strong say in shaping energy policies and prices. In order to prevent oil prices from falling, OPEC members have always sought to keep their production at a certain level. Because it is not a member of OPEC, Russia is not under any obligation to cut production or adhere to a quota system. Thus, in 2001, OPEC cut its output by 13 percent, while Moscow increased its production and share of global market. This imbalance prompted OPEC to seek Russia's cooperation to prevent prices from further falling. After some hesitation, Russia agreed to cut production and work with OPEC to restore stability in oil prices and markets.

Third, the September 11 terrorist attacks and the subsequent war on terrorism have strengthened the argument for reducing Western dependence on the volatile Middle East. (Fifteen of the hijackers are believed to have come from Saudi Arabia.) This has given Russia a great opportunity to present itself as a reliable energy partner to the United States and other Western countries with massive and secure resources.

Finally, this current and projected close cooperation between Russia and the West should not be exaggerated. The rivalry between Moscow, Brussels, and Washington will continue in the Persian Gulf, East Asia, and the rest of the world. This rivalry, however, is increasingly driven by economic interests and less by ideological or nationalistic sentiments.

Pacific Asia

Since the early 1990s the Pacific Asia region—Australia, Bangladesh, China, India, Indonesia, Japan, Malaysia, New Zealand, Pakistan, Philippines, Singapore, South Korea, Taiwan, and Thailand—has rapidly emerged as a central player on the global energy scene. This can be explained by the fundamental changes in the region's pattern of energy consumption, its current and projected skyrocketing demand, and heavy dependence on foreign suppliers.

First, a close examination of Pacific Asia's energy mix indicates a serious imbalance between its indigenous hydrocarbon resources and its growing demand. Table 2.2 shows how poor the region is in hydrocarbon resources (with the exception of coal), particularly when the region's population is taken into account. One third of the world's population holds a small fraction of proven oil and gas reserves. Also, the table shows the region's low level of oil consumption, particularly when compared with the United States. (About 280 million people, approximately 5 percent of the world's population, consume more than 25 percent of world's oil.) The gap is far larger when comparing natural gas consumption. This means that per capita oil and gas consumption in Pacific Asia is much lower than in developed countries, and the high economic growth in several Asian states for the past several years will narrow this gap.[20]

Second, most of the region has achieved a high level of economic growth, unparalleled in the rest of the world. As Asian economies rocketed, their energy consumption grew by similar proportion. The region's economic expansion was the engine driving higher global oil demand. In the early 1990s total energy demand grew at an average rate of 5.5 percent per year, compared with the world average of 1.5 percent per year.[21] Indeed, between 1988 and 1997, the region's energy demand eclipsed Europe's and became second only to that of the United States.[22] In mid-1997, however, the region experienced a severe financial and economic crisis with significant impact on worldwide energy demand and prices. The economic downturn started in Thailand and quickly spread across Southeast Asia. Asian economies were not equally affected by the crisis, with Thailand, Indonesia, and South Korea experiencing the most severe initial shocks. Hong Kong was also greatly affected due to its role as a regional trade hub. China and Taiwan experienced relatively mild effects, while the Philippines and Malaysia were somewhat less severely affected. The Asian economic crisis contributed to the sharp decline in world oil prices from late 1997 through early 1999. The initial projections that East Asia's re-

Table 2.2. Fossil fuels in Pacific Asia in 2000: Proven reserves, production, and consumption as a percentage of total world

	Oil	Natural gas	Coal
Proven reserves	4.2	6.8	29.7
Production	10.6	11.0	43.3
Consumption	27.8	12.1	43.4

Source: BP/Amoco, *BP Statistical Review of World Energy* (London, June 2001).

covery from its economic crisis would be slow in coming seem to have been too pessimistic, at least in most cases. Today, most Pacific Asian economies have resumed their growth and demand for energy.

Third, given the proximity of the Persian Gulf oil and gas producers to the vast and growing market in Pacific Asia, the two regions have founded a de facto energy partnership. For the past several years East Asian states have increasingly grown dependent on oil and gas supplies from the Persian Gulf states. This dependency is expected to deepen over the next few decades.[23]

To sum up, the expected dramatic increase in oil and gas consumption in Asia and the anticipated growing role of the Persian Gulf and the Caspian Sea producers in satisfying this demand are likely to have a long-term significant impact on the global energy scene as well as on the geopolitics of these regions. A close examination of China and Japan would shed light on these likely changes.

China

Since the 1980s, the Chinese economy has grown faster than almost any other country. Naturally, high economic growth requires more energy consumption. Thus, China has become the second largest energy-consuming nation, trailing only the United States. This skyrocketing demand for energy has raised concern over the adequacy of indigenous resources to satisfy the ever-growing consumption. In other words, security of energy supply is considered crucial for sustaining China's economic development. The problem is that China's oil production has not kept pace with its growing demand since the early 1990s.

In order to address this imbalance, Beijing has pursued a threefold strategy. First, in 1993 China shifted away from self-sufficiency and became a net oil importer. Second, Chinese companies have been encouraged to acquire interests in oil and gas exploration and production operations abroad, particularly in places where the Chinese government has political leverage such as Iran, Iraq, and Central Asia. Third, Beijing has embarked on an ambitious and comprehensive plan to reform its oil industry and invite foreign investment. In 1998, the Chinese government reorganized most state-owned oil and gas assets into two firms, the China National Petroleum Corporation and the China Petrochemical Corporation.[24] Meanwhile, foreign oil companies (including British Petroleum, Chevron, Enron, and Japan National Oil Company) started exploring for oil in onshore and offshore blocks.[25]

For much of its modern history, China has relied on coal and oil to fuel its economic growth. Natural gas has played only a marginal role in satisfying the country's growing energy needs. However, according to many environmental groups, China is one of the most polluted countries in the world, and a new awareness is emerging that clean energy sources must be utilized to maintain a sustainable social and economic development. Thus, a significant shift is emerging in China's energy mix. Today, gas accounts for only slightly more than 3 percent of total energy mix, but consumption is expected to more than triple by 2010.[26] A major obstacle to the large-scale development of the gas industry is the geographical imbalance between gas reservoirs and consuming centers. The majority of gas reserves are located in the far western region of Xinjiang, and there is no national grid to link them with the main population centers in the east. A massive investment in pipeline infrastructure needs to be made.

In addition to utilizing its domestic gas resources, Beijing has sought to import natural gas from abroad. Overseas priority has been given to neighboring hydrocarbon provinces including the Persian Gulf, Russia, and Central and Southeast Asia. Several countries including Russia, Kazakhstan, Turkmenistan, Australia, Indonesia, Malaysia, Qatar, Yemen, and Iran have negotiated natural gas agreements with the authority in Beijing. Still, costly international pipeline links and coastal LNG receiving facilities must be put in place to make such imports viable.

Since the early 1990s, a growing gap between China's oil and gas production and consumption has emerged. The Chinese leaders have realized that in order to sustain their country's impressive high economic growth, they have to secure adequate and secure supplies. This goal can be achieved by inviting foreign investment to fully utilize the country's indigenous resources and building economic and strategic partnerships with current and potential oil and gas suppliers. Put differently, a growing interdependence relationship has been forged between Western and Russian companies and governments and their Chinese counterparts. This cooperative relationship has another side: rivalry. The two sides compete with each other over oil and gas deals in the Persian Gulf, the Caspian Sea, and other hydrocarbon provinces. Chinese companies are involved in oil and gas industries in Iran and Iraq, and for a long time Beijing has established friendly political and military relations with both Tehran and Baghdad. Similarly, in November 1999 President Jiang Zemin made the first-ever visit by a Chinese head of state to Saudi Arabia. In short, China's foreign policy in the Persian Gulf and central and eastern Asia is increasingly driven more by geo-economic interests and less by geostrategic concerns.[27]

Japan

Japan is a significant player on the international energy scene. Its gross domestic product is the second largest, after the United States. The Japanese economy, however, experienced a recession for most of the 1990s. The country's energy sector is different in many ways from other consumers.

First, Japan has almost no hydrocarbon reserves of its own but is the world's fourth largest energy consumer (after the United States, China, and Russia). Consequently, Japan imports over 80 percent of its fuel, the highest percentage of any major industrialized nation.[28] As a result of slow economic growth, Japanese demand for energy has been stagnant in recent years.

Second, since the first oil shock in the mid-1970s, a driving factor shaping Japanese energy policy has been fear of an oil disruption. Thus, Tokyo has sought and succeeded in reducing its heavy dependence on oil. Between 1973 and 2000, oil's share in Japan's total primary energy consumption fell from 77 percent to 55 percent.[29] Three policies have been employed to achieve this success: (1) Japan promoted energy efficiency, and now its energy intensity (energy use per unit of GDP) is among the lowest in the developed world; (2) the Japanese economy shifted from heavy industry toward service; and (3) substantial investments were made to promote energy diversification and reliance on nuclear power and natural gas instead of oil.

Nuclear power was long considered the best source of energy because it is domestically produced and thereby more secure than other sources. Little wonder, Japan's nuclear output nearly doubled between 1985 and 1996, and Tokyo has emerged as the third country in the world with installed nuclear capacity (behind the United States and France).[30] The future of nuclear power in Japan, however, does not look promising. Public outcry over recent incidents involving the dangerous mishandling of nuclear materials has forced the Japanese government to alter its ambitious nuclear energy plans. On the other hand, natural gas provides a growing share of Japan's energy needs. Indeed, Japan has established itself as a major gas-consuming nation. Most of this gas is in the form of LNG and comes mainly from southeast Asia (particularly from Indonesia and Malaysia).

Third, despite oil supply interruptions in the mid-1970s and early 1980s, Japan has not succeeded in reducing its dependence on the Middle East. Most of Tokyo's imported oil continues to come from Persian Gulf states.

These peculiar characteristics of Japan's energy stand have shaped foreign policy. Since the end of World War II, Japan has sought to concentrate on economic targets, leaving security to the United States and showing reluctance to become involved in international conflicts. Consequently, despite heavy dependence on imported oil and gas, Tokyo does not have the necessary military means to protect its energy and economic interests. Rather, when there is a threat to oil supplies (e.g., when Iraq invaded and occupied Kuwait in August 1990), Japan relies on the United States to ensure the security of oil shipments. Lacking military capability to protect its strategic and economic interests, Japan has sought to employ diplomacy and investment to influence developments in major energy-producing countries. By fostering economic ties and strengthening the economies of these states, Japan hopes to provide the foundation to future democratic development. Japan's huge volume of trade with Saudi Arabia and the other Gulf monarchies is a good illustration of this policy. Similarly, Japan expressed its opposition to U.S. economic sanctions against Iran, although Tokyo reluctantly observed them under American pressure. Still, Japan has resumed its policy of providing loans to Iran, and Japanese firms have been granted priority negotiating rights to develop the recently discovered Azadegan oil field in Iran.

This discussion of the main characteristics of the energy industry in the European Union, Russia, and Pacific Asia underscores a significant development. Despite deep-rooted conflicting strategic interests, they all share the same goal: the stability of energy global markets. Energy consumers and producers work together to ensure adequate investment, appropriate technology, proven reserves, and secure supplies. This cooperation/rivalry between several international players is also important in containing and reducing global pollution.

Global Warming

The earth's atmosphere acts as a filter for solar rays. Half of the visible light and ultraviolet radiation given off by the sun is either absorbed by various layers or reflected back into space. Most of the light that does get through heats the earth's surface and is eventually reflected back into space as infrared radiation. The "greenhouse effect" is the atmospheric trapping of that infrared radiation, a natural phenomenon without which the earth would be uninhabitably cold for humans. During the combustion of carbon-based fossil fuels, greenhouse gases such as carbon dioxide, methane, and nitrous oxide are emitted. These gases add to that atmospheric layer which is permeable to ultraviolet but not infrared radiation. As more fossil fuels are burned, the layer of greenhouse gases thickens; solar radiation continues to pass through unimpeded, while heat reflected from the earth finds it harder and harder to escape into space. This causes the gradual increase in the earth's temperature known as global warming.[31]

Recent international efforts to protect the environment took a significant turn in December 1990 when the United Nations created the Framework Convention on Climate Change (UNFCCC). This was adopted and opened for signature at the UN Conference on Environment and Development, the "Earth Summit," in Rio de Janeiro in 1992. Participants committed themselves to stabilizing greenhouse gas concentrations and preventing harmful interference with the climate system. The Convention further adopted a list of industrialized nations (Annex I) for which domestic/international greenhouse gas reduction measures were recommended and developing nations (non–Annex I) that are exempt from immediate emission reduction measures but may participate on a voluntary basis. Since the UNFCCC entered into force in 1994, the parties have continued to negotiate an agreement to achieve their goals. These negotiations have mainly taken place in the so-called Conference of the Parties (COP). Subsequent COPs have been held annually. The Kyoto Protocol was adopted at COP 3 in December 1997. It left many of its operational details unresolved, however. These details were further negotiated, and in COP 7, held in Marrakech, Morocco, in November 2001, an agreement was reached, enforcing the Kyoto Protocol.

Mechanism

The Kyoto Protocol represents a potentially binding international treaty that stipulates actions to be taken by nations to combat global climate change.³² Most industrialized countries have agreed to reduce their collective emissions of greenhouse gases by an average of 5.2 percent below levels measured in 1990 by the commitment period of 2008–12.³³ The Protocol provides six options for meeting emissions reduction targets. The first two are identified as national approaches, while the last four are identified as international ones.

> *Policies and measures:* These are actions taken directly by government agencies that have the effect of reducing emissions (e.g., the adoption and enforcement of new laws and regulations).
> *Sink enhancement:* A carbon sink is a reservoir that can absorb carbon dioxide from the atmosphere and include forests, soils, peat, permafrost, and ocean water. The Protocol allows Annex I parties to earn credits for activities that increase the natural removal of carbon dioxide from the atmosphere through the enhancement of sinks.
> *Joint fulfillment:* This allows for burden sharing among the fifteen EU countries so that not all of them will necessarily reach the 8 percent target. Similarly, any other group of Annex I parties is also entitled to enter into a "joint fulfillment" agreement.
> *Joint implementation:* The Protocol provides for credits from emissions reduction projects to be transferred between Annex I parties.
> *Clean development mechanism:* The Protocol grants Annex I parties the right to generate or purchase emissions reduction credits from projects undertaken by them within non–Annex I countries. In exchange, developing country parties will have access to resources and technology to assist in the development of their economies.
> *Emissions trading:* Annex I parties that have reduced their emissions below their allowances will be able to trade some part of the surplus allowances to other Annex I parties.³⁴

Reactions

Governments from all over the world have reacted to the Kyoto Protocol based on their perceived national interests and the level of their economic development. The European Union generally supported the Protocol because it believes it has a good chance of meeting the targets without significant detriment to European economies. From 1990 to 1999, economic growth in the EU averaged 2 percent, while the growth in carbon emissions was only 0.16 percent, making Europe the only continent to be emitting less carbon in 1999 than in 1990.³⁵ For Japan, the 6 percent reduction target was a significant move from its initial proposal of a 2 percent cut in emissions. The Japanese government decided to rely on both national (e.g., domestic programs) and international approaches (e.g., flexibility mechanisms) to achieve this goal.[36]

On the other hand, many developing countries, led by China, have contended that, because rising atmospheric greenhouse gas concentrations were the result of past economic development by developed countries, it was proper for those same countries to lead the way toward a solution. According to this argument, it is hypocritical of developed countries that produce the most pollution and use the most resources to expect the poor to make sacrifices.

Given the leading role the United States plays on the global scene as the world's largest economy, energy consumer, and polluter, the American stand on the Kyoto Protocol is particularly significant.[37] From the beginning, it was evident that the United States would have the most difficulty meeting its target due to its high energy per capita consumption. Thus, in 1997 the Senate adopted a nonbinding resolution that opposed the essential features of the Kyoto Protocol. The Clinton administration largely supported the overall goal of trying to reduce greenhouse gas emissions but was concerned about the potential effect radical reductions might have on the U.S. economy. In an attempt to hold the projected cost of mandated reductions to a level that the economy could afford, the Clinton administration tried to negotiate an arrangement that would have permitted economic efficiency to play a major role in the emissions reduction process by making use of market mechanisms and greenhouse gas sinks.[38] President Clinton never submitted the Protocol to the Senate for ratification.

George W. Bush called the Kyoto Protocol "fatally flawed," and in March 2001, just weeks after taking office, he declared that he had no interest in implementing or ratifying the treaty. In other words, the United States withdrew from the Kyoto Protocol. According to the Bush administration, the Kyoto Protocol risks significantly harming the American economy. The U.S. Energy Department projected that the implementation of the treaty would reduce our GDP by 1–2 percent by 2010.[39] Second, the Kyoto Protocol's targets are not based on science. The targets and timetables were arrived at arbitrarily as a result of political negotiations, not sound scientific evidence. Third, it excludes developing countries. These countries' net emissions are rapidly increasing and rivaling those of developed countries. Still, following the withdrawal from the Kyoto Protocol, Bush created a task force to come up with alternatives.

Saudi Arabia and other producers and exporters of hydrocarbon resources share similar concerns to those expressed by the United States regarding the Kyoto Protocol. Their main concern is less about curtailing greenhouse gases and more about preventing economic disruption. A worldwide reduction in fossil fuel consumption, as negotiated in the Kyoto Protocol, would certainly lead to substantial export revenue losses for oil producers.[40] In order to counter the Kyoto Protocol's potential devastating impact on their economies, Saudi Arabia and other oil producers have requested that the developed countries consider the following steps: (1) assisting heavily dependent fossil fuel exporting countries with investment and transfer of technologies to enable them to achieve diversification and reduce dependence on oil; (2) compensating fossil fuel exporting countries for any proved economic injuries resulting from mitigation measures; (3) removing all market distortions such as subsidies and tax incentives that would result in encouraging more production of fossil fuels in the developed world.[41]

These different reactions to the Kyoto Protocol raise concern over the direction the international community is likely to take in addressing the issue of global warming. There is no doubt that the Kyoto Protocol is a historic agreement, but it is not the final solution. President Bush's decision not to seek ratification of the Protocol has thrown uncertainty into the process. While there is agreement that cutting air pollution is a worthwhile goal in itself and that emissions reductions would benefit from international cooperation, disagreements arise as to what should actually be done. Reconciling U.S. concerns and international cooperation is also essential in the full utilization of hydrocarbon resources in the Persian Gulf and the Caspian Sea.

3

Managing Dependence

American-Saudi Oil Diplomacy

Very few relations can rival those between the United States and Saudi Arabia, which have been built on long-standing traditions of friendship and informal alliance. Oil is at the heart of this close American-Saudi cooperation. Leaders on both sides perceive the substantial and uninterrupted oil supplies from the kingdom as one of the most important elements of global energy security and development of world economy. Thus, for more than half a century, Saudi Arabia, in cooperation with the U.S. government and American oil companies, has played a pivotal role in promoting moderation and stability in energy pricing and policies.

Unlike other major producers in the Persian Gulf and the Organization of Petroleum Exporting Countries (OPEC), oil explorations and developments in Saudi Arabia have been carried out almost entirely by American companies. In the early 1930s, U.S. oil companies were looking for commercial opportunities overseas. Promising oil reservoirs had been discovered in Iran, Iraq, and Bahrain. This newly discovered hydrocarbon wealth was dominated by European companies, particularly from Great Britain. Meanwhile, indigenous leaders were interested in granting concessions to foreign companies in order to strengthen their rising economic and political power. Under these circumstances, in 1933 King Saud Ibn Abd al-Aziz, the founder of modern-day Saudi Arabia, who was suspicious of the European intentions, gave Standard Oil Company of California (Socal, later Chevron) a sixty-year exclusive right to explore for oil in an area in eastern Saudi Arabia covering 360,000 square miles.[1] The Californian-Arabian Standard Oil Company (CASOC) was formed to exploit the concession. In 1936, Socal sold a half share in CASOC to the Texas Oil Company (Texaco), and later two more American companies acquired shares: Standard Oil Company of New Jersey (later Exxon) and Standard Oil Company of New York (originally Socony, later Mobil).[2] A supple-

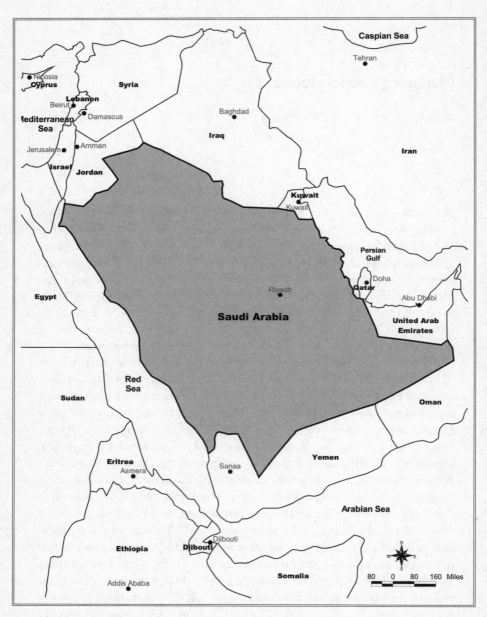

Map 3. Saudi Arabia

mentary agreement was signed in May 1938, adding six years to the original agreement and enlarging the concession area by almost 80,000 square miles. It also included rights in the Saudi government's half-interest in the two neutral zones shared with Iraq and Kuwait.

Early exploration drilling in Saudi Arabia was not successful, and although the first well was completed in 1935, it was not until March 1938 that oil was struck in commercial quantities in the Dammam structure. First exports of oil took place in 1938 and continued at very modest levels until after World War II. But the event that transformed prospects for the oil industry in Saudi Arabia was undoubtedly the discovery of the Ghawar field in 1948, which proved to be the world's largest single oil-bearing structure.[3] The world's largest offshore field, Safaniya, lies in Saudi Arabian waters of the Persian Gulf. In 1944, CASOC was renamed the Arabian American Oil Company (Aramco).

Unlike other foreign oil companies, Aramco had good relations with the host government, Saudi Arabia, and with the local population. The bitter dispute in the early 1950s between the Iranian authority and British Petroleum was very different from the smooth cooperation between Aramco and the Saudi government. In 1950 the Saudi government and Aramco reached an agreement on a modified system of profit-sharing, which introduced the notion of the 50/50 division between the host country and the concessionaire. In 1973, the Saudi government took a 25 percent stake in Aramco. A year later, this share was increased to 60 percent, and in 1980 it was amicably agreed that Aramco should become 100 percent Saudi-owned, with the date of ownership backdated to 1976.[4] Prior to the Saudi takeover, Aramco had been the largest single American investment in any foreign country. This friendly and nonconfrontational change of ownership helped the two sides to maintain their cordial cooperation. Despite the Saudi takeover of Aramco, U.S. administrators and technicians, side by side with their Saudi counterparts, continued to occupy important positions in the company. Finally, in April 1989, the last American to preside over Aramco, John J. Kelberer, handed over power to its first Saudi boss, Ali al-Naimi, who later became oil minister.

To sum up, Aramco had a great success story in terms of profits and relations with the local population and the host government. This success story illustrates the mutual understanding between Saudi Arabia and the United States of the significance of their partnership and cooperation in promoting stability of global energy markets and world economy. The next sections examine two crucial areas in the American-Saudi relations:

oil pricing and the reopening of Saudi energy sector to foreign investment. Finally, I address some of the main challenges facing the two countries in their efforts to stabilize energy markets, particularly security, economic and political reforms in the kingdom, the impact of the Arab-Israeli conflict/peace process on the bilateral American-Saudi relations, and militant Islam and the war on terrorism since September 11, 2001.

Oil Prices

The price of oil is critical to world economy, given that oil is the largest internationally traded commodity, both in volume and value.[5] To explain the impact of price fluctuations on oil producers, consumers, and the global economy, I will discuss three points: price benchmarks, the ups and downs of oil prices, and the volatility that has characterized oil markets since 1997. I believe the question of which is worse, high prices or low prices, is debatable. There are advantages and disadvantages for both. Rather, the volatility of the swings is the most dangerous to energy markets and global economy.

Reference Prices

Oil producers use several crude blends in their trade with oil consumers. OPEC collects pricing data on a "basket" of seven crude oils (Algeria's Saharan Blend, Indonesia's Minas, Nigeria's Bonny Light, Saudi Arabia's Arab Light, Dubai's Fateh, Venezuela's Tia Juana Light, and Mexico's Isthmus—a non-OPEC crude oil) to monitor world oil market conditions. However, the OPEC basket is used for statistical purposes and not for the actual buying and selling of oil. Instead, the two most actively traded crude oil blends are North Sea Brent and U.S. West Texas Intermediate (WTI). Brent is traded on the International Petroleum Exchange in London, while WTI is traded on the New York Mercantile Exchange. Both crude oils also are traded on spot markets.

Brent is the principal grade of the United Kingdom North Sea crude oil in international oil trading, originating from the Brent and other fields of the East Shetland Basin.[6] It is used as the "marker" for other North Sea grades, which trade at differentials to it reflecting quality and location. Brent is also lighter, sweeter, and more expensive than the OPEC basket, although less so than WTI. Because WTI is a very light, sweet (low sulfur content) crude, it is more expensive than the average OPEC basket. Saudi Aramco's use of WTI for deliveries to North America is based on the desire to use a well-traded crude. WTI is confined to one main inland pipeline system connecting Texas with the Chicago–Great Lakes refining area. However, it is widely accepted as the basis for pricing most American and Canadian crude oil.

For sales to Asia, Saudi Aramco, like most other Middle Eastern oil exporters, bases its prices on those of Dubai's Fateh crude. While Dubai attracts a good cross-section of traders, its production is in long-term decline. Thus some exporters, including Saudi Arabia, have added a second benchmark crude, Oman, to their export price formulae. The government of Oman is anxious to avoid the situation where it becomes the sole benchmark for Middle Eastern crude. It fears that this will make its own export prices more volatile.[7] Finally, it is important to point out that since April 2000 Saudi Aramco has changed its method of determining prices for long-term contract sales to Europe. Instead of relying on Brent spot market prices to determine the price for its oil to European customers, it started using a "B-Wave."[8] The "B-Wave" uses a weighted average of Brent futures prices, which is less volatile than the spot market price.[9]

To sum up, Saudi Arabia, like other oil producers, uses different reference prices in exporting its crude to consumers. Saudi Aramco has sought to introduce some changes to its price formulas in order to bring about stability and consistency in oil prices and markets.

Background

When petroleum burst onto the world stage in 1859, its price first went through some initial gyrations (1860–70), before settling in the $1–$2 per barrel range (barring a few exceptions) for a full century.[10] Since the early 1970s, however, volatility has been the main theme of oil prices and markets.

The figures in table 3.1 show the strong impact of political developments in the Middle East on the stability of global prices. From the early 1970s to the early 1990s, there were the so-called three price shocks and a reverse shock. The first price shock followed the 1973 Yom Kippur War between the Arabs and Israel. The second was in response to the Iranian revolution of 1979 and the beginning of the eight-year war between Iran and Iraq (1980–88). The third shock was a short one following the Iraqi invasion of Kuwait in 1990. In each case, the threat to a stable future supply of Middle East oil exports led to a sudden increase in the price of oil. The reverse wave in 1986 reflected a major change in the Saudi oil policy.

Table 3.1. Crude prices in $U.S. per barrel, 1972–2000

Year	Dubai	Brent	WTI
1972	1.90	—	—
1973	2.83	—	—
1974	10.41	—	—
1975	10.70	—	—
1976	11.63	12.80	12.23
1977	12.38	13.92	14.22
1978	13.03	14.02	14.55
1979	29.75	31.61	25.08
1980	35.69	36.83	37.96
1981	34.32	35.93	36.08
1982	31.80	32.97	33.65
1983	28.78	29.55	30.30
1984	28.07	28.66	29.39
1985	27.53	27.51	27.99
1986	12.95	14.38	15.04
1987	16.92	18.42	19.19
1988	13.19	14.96	15.97
1989	15.68	18.20	19.68
1990	20.50	23.81	24.52
1991	16.56	20.05	21.54
1992	17.21	19.37	20.57
1993	14.90	17.07	18.45
1994	14.76	15.98	17.21
1995	16.09	17.18	18.42
1996	18.56	20.81	22.16
1997	18.13	19.30	20.61
1998	12.16	13.11	14.39
1999	17.30	18.25	19.31
2000	26.24	28.98	30.37

Source: BP/Amoco, *BP Statistical Review of World Energy* (London, 2001), 14.

For a long time, the idea of using oil as a political weapon by Arab producers against Western powers that support Israel has been considered. Thus, in the Suez crisis (1956) and the Six-Day War (1967), limited embargoes were implemented with little effect. In 1956, the United States was able to compensate for the embargo against Britain and France from its own domestic production. In 1967, Saudi government ordered Aramco to stop oil supplies to the United States and the United Kingdom during the Arab-Israeli war. The ability of the international oil companies to reroute supplies, however, made the embargo ineffective. Furthermore, the Saudi leaders were not fully convinced of the validity of mixing oil with policy. Rather, the Saudi leaders preferred to use oil revenues as a "positive weapon" to build up the military and economic strength of the Arab world.[11] Accordingly, Saudi Arabia and other Arab oil producers gave substantial financial aid to Egypt and Jordan from 1967 to 1973.

Debate about the potential of the oil weapon continued nonetheless. Shortly after the outbreak of the Yom Kippur war in 1973, President Nixon asked the Congress to provide $2.2 billion in emergency security assistance to replace Israel's losses in the war.[12] On the following day Saudi Arabia, which had not gone beyond cutting production and issuing warnings during the fighting up to that point, announced a total embargo on oil shipments to the United States.[13] Other Arab states on the Persian Gulf followed suit. In the following months Saudi production dropped sharply creating circumstances that led to a skyrocketing in the posted price of oil. It was this cutback in production, not the embargo, that produced the shortages felt for several months in most industrialized countries.

The embargo was lifted in March 1974 after Secretary of State Henry Kissinger had embarked on the "shuttle diplomacy" that had brought disengagement agreements between the Israeli troops on one side and the Egyptian and Syrian armies on the other side. This probably was one of the lowest points in the relations between Washington and Riyadh. Another sign of these deteriorated relations was Kissinger's warning that the United States would not rule out the use of military force to secure oil supplies from the Persian Gulf region.[14] Saudi Arabia and the United States were able to overcome this short-lived crisis in their relations. Under the Carter administration, the talk about potential use of military force ceased and, instead, the administration proposed and the Congress approved the sale of some of the most advanced weapons in the U.S. arsenal to Saudi Arabia.

It is also important to point out that in response to this first oil shock, representatives of the Organization for Economic Cooperation and Development (OECD) convened a conference in Washington, D.C., in 1974 that led to the establishment of the International Energy Agency (IEA). The proposal to set up an energy organization was made by Kissinger and accepted by most members of the OECD.[15] The emphasis was on supply security. In order to improve energy security, the participating countries pledged to hold oil stocks equivalent to ninety days of net imports. They also developed an integrated set of emergency response measures that included demand restraint, fuel switching, and surge oil production. These measures also included the important provision of stock drawdown and sharing of available supplies in the event of oil supply disruptions involving a loss of 7 percent or more for any member country or for the group.[16]

In short, the IEA was created by oil-consuming countries to coordinate their efforts against those made by oil-producing countries, members of OPEC. Since then, these confrontational attitudes adopted by the IEA and OPEC have waned, and cooperation to ensure global energy stability has characterized the relations between the two organizations.

The second oil shock (1979), like the first one, was the outcome of a complex interaction of political and economic developments. The domestic political turmoil that led to the Iranian revolution resulted in the withdrawal of most Iranian oil production from the international market. In their efforts to undermine the Pahlavi regime the Iranian oil workers went on strike, bringing the country's oil production almost to a complete halt. Initially, the gap was filled by Saudi Arabia. However, a few months later, the Saudis cut back production. Historians debate whether this decision was the result of technical difficulties in maintaining production so close to capacity or of Saudi dissatisfaction with the U.S.-brokered Camp David Agreement between Egypt and Israel.[17]

It was under these circumstances that Saudi Arabia, to the shocked disappointment of American officials, adhered to the tough sanctions against Egypt proposed by other Arabs, who opposed the unilateral Egyptian move to make peace with Israel.[18] The oil-producing countries, including Saudi Arabia, also agreed to provide substantial new annual subsidies to the remaining "confrontation" states—Jordan and Syria—as well as to the Palestine Liberation Organization. These developments in the late 1970s and early 1980s—the Iranian revolution, the Iran-Iraq war, and Saudi reluctance to increase production—pushed oil prices to their highest level ever.

These extremely high prices did not last long. In the mid-1980s oil prices fell significantly. At least three factors contributed to this drop. First, Saudi Arabia decided to abandon the role of swing producer. Second, Riyadh sought to reclaim its share of the global oil market. Third, the kingdom wanted to force Iran to end the war with Iraq. The sudden and unprecedented rise in oil prices of the second half of the 1970s brought a high level of uncertainty and instability in oil markets. In an attempt to "manage supplies" and restore a sense of predictability and consistency in the global markets, OPEC introduced the quota system in 1982, under which all OPEC members agreed to limit their production to a certain level. The high prices, however, meant that there was a great opportunity to make substantial profits, and subsequently there was very little compliance. Most OPEC members produced more than their stated quota. Being the largest producer in OPEC, Saudi Arabia was trying to maintain a sense of coherence in the organization by adjusting its level of production to those of other members. In 1985–86, Riyadh changed course and flooded the market with its own cheap oil.

Meanwhile, high oil prices and political and economic instability in the Middle East convinced major oil-consuming countries to reduce their dependence on oil imported from the Persian Gulf by consuming less oil (conservation), investing in other sources of energy (natural gas, nuclear power, and renewable resources), and increasing exploration and production operations outside the Persian Gulf (most prominently in the North Sea). Together, these steps resulted in a substantial decline in Saudi Arabia and other Persian Gulf producers' share of the global market. Put differently, the share of producers in the North Sea, Latin America, and other regions rose at the expense of those in the Persian Gulf. Saudi Arabia wanted to reverse this trend. Since the mid-1980s, Saudi Arabia and other Persian Gulf producers' share in the oil global market has been on the rise.

Finally, during the Iran-Iraq war, Riyadh supported Baghdad against Tehran. In the mid-1980s, the Iranian army was engaged in a major military push inside Iraq. By substantially reducing the price of oil, the main source of national income to Iran that provided revenues to sustain war efforts, Riyadh dealt a heavy blow to the Iranian war efforts

This deliberate Saudi policy to bring oil prices down ended the oil boom. For the following decade (1987–97) prices ranged between $15 and $20, a level with which both consumers and producers seemed happy enough, despite some considerable volatility around the mean. Most notably, in 1990 the Iraqi invasion of Kuwait caused a temporary and short-lived shortage of supplies, which caused panic in the global markets and pushed prices up. In a very short time, Saudi Arabia increased its production to make up for the Iraqi and Kuwaiti supplies, and prices fell to their average prior to the crisis. Developments in the late 1990s, however, introduced new elements to the oil markets and brought back a sense of volatility and uncertainty to the markets and world economy.

Volatility since 1997

From 1997 to 2000, crude oil prices were more volatile than at any time in recent history. In a matter of a few months, prices hit their highest and lowest levels since the price collapse of the mid-1980s. OPEC played a crucial role in this stunning fluctuation. This can be explained by the organization's huge share of global oil production and its share of the internationally traded oil. Furthermore, OPEC members have increasingly worked with other major oil producers particularly in Mexico, Russia,

Table 3.2. Saudi oil production and share of total OPEC, 1970–2000 (thousand barrels per day)

Year	Saudi production	OPEC production	Saudi as % of OPEC
1970	3,799	23,301	16.30
1971	4,769	25,209	18.91
1972	6,016	26,891	22.37
1973	7,596	30,629	24.80
1974	8,480	30,351	27.93
1975	7,075	26,771	26.42
1976	8,577	30,327	28.28
1977	9,245	30,893	29.92
1978	8,301	29,464	28.17
1979	9,532	30,581	31.16
1980	9,900	26,606	37.20
1981	9,815	22,481	43.65
1982	6,483	18,778	34.52
1983	5,086	17,497	29.06
1984	4,663	17,442	26.73
1985	3,388	16,181	20.93
1986	4,870	18,275	26.64
1987	4,265	18,517	23.03
1988	5,086	20,324	25.02
1989	5,064	22,071	22.94
1990	6,410	23,195	27.63
1991	8,115	23,375	34.71
1992	8,332	24,398	34.15
1993	8,198	25,119	32.63
1994	8,120	25,510	31.83
1995	8,231	26,004	31.65
1996	8,218	26,461	31.05
1997	8,362	27,710	30.17
1998	8,389	28,774	29.15
1999	7,833	27,579	28.40
2000	8,404	29,113	28.86

Source: Energy Information Administration, *International Petroleum Monthly* (Washington, D.C.: Government Printing Office, June 2001), 38.

and Norway to reach a consensus on a common oil policy that would benefit all participants. Saudi Arabia is the unofficial leader of OPEC. This is due to the kingdom's substantial share of OPEC production.

The figures in table 3.2 show not only the huge Saudi production both in absolute terms and as a percentage of total OPEC production. They also show that Saudi Arabia's share in total OPEC production reached its highest levels in 1981 and 1991. In other words, during the Iran-Iraq war and the Gulf war, Riyadh increased production to make up for the shortage of supplies caused by regional conflicts. This Saudi pivotal role in OPEC and global oil markets can be attributed to massive reserves, low production costs, and spare capacity.

With 261 billion barrels of proven oil reserves (more than one fourth of the world total) and up to 1 trillion barrels of ultimately recoverable oil, Saudi Arabia is the world's leading producer and exporter.[19] In addition to this enormous oil reserve base, Saudi Arabia's cost of production is one of the lowest in the world. The kingdom's cost of production is less than $1.50 per barrel, while the global average cost is about $5 per barrel and the costs in some places are even higher. Also, Saudi Arabia has a great advantage when it comes to adding new reserves or increasing production capacity. According to the Saudi oil minister, Ali al-Naimi, it costs the kingdom less than 10 cents a barrel to discover new reserves, while the cost in some other areas of the world can be as high as $4 per barrel.[20] In short, Saudi current production costs and the costs for developing more production capacity for the future are probably the lowest in the world. Finally, most of the world's spare productive capacity is located in Saudi Arabia.[21] This is an important strategic asset for the kingdom. It means that whenever a sudden interruption of supplies occurs, Saudi Arabia can quickly fill the gap. This can be seen as an insurance policy against temporary shortages in world oil supplies. Riyadh has proven, in most cases, a reliable partner in using its spare production capacity to overcome oil crises. Given all these characteristics of the Saudi oil resources, the kingdom is a critical player in determining global oil prices and the stability of energy markets. As such, Saudi Arabia was an important player behind the oil price collapse of late 1997 through early 1999 and also in actions taken by world oil producers, which have led to a tripling in oil prices in 1999–2000. Finally, in the aftermath of the September 11 terrorist attacks, Riyadh and other oil producers have sought to calm the market and restore a sense of stability.

To understand these strong swings in oil prices in the late 1990s and early 2000s, it is important to consider three factors that influence Saudi policy. First, a decline in prices benefits oil-consuming countries. It acts similar to a tax cut, increasing consumer disposable income. This allows for looser monetary policy and hence lower interest rates with lower inflation and stronger economic growth than would otherwise be the case. On the other hand, sharply higher oil prices were a major cause for recessions in the United States and other Western economies for the last several decades and are believed to have contributed to the slowdown of the American economy in the early 2000s.[22]

Second, for Saudi Arabia and other Persian Gulf oil producers, very high prices encourage exploration and development investments in other regions such as the Caspian Basin, the North Sea, and the Gulf of Mexico. Yet excessively low prices also have harmful political and economic impacts. Considering that oil revenues are the main source of national income in Saudi Arabia and other oil producers, very low prices mean shrinking financial resources and reduced public spending.

Third, in 1960 Saudi Arabia joined Iran, Iraq, Kuwait, and Venezuela in creating OPEC. Since then it has been the kingdom's preference, where possible, to work within the OPEC decision-making structure. As a founder and a leader of OPEC, Riyadh has always sought to reconcile its national interests with those of fellow members and has been engaged in consensus-building not only among OPEC members but also between OPEC and other major oil producers. In spite of these efforts by Saudi Arabia and other oil producers to reach a consensus and present a coherent and unified front, OPEC does not always speak with one voice. Traditionally, the organization's market power has been viewed as a trade-off between maximizing price and maximizing market share. Within OPEC, countries like Algeria, Nigeria, and Indonesia, for instance, contain relatively large populations and relatively small oil reserves. These countries, therefore, have tended (with numerous exceptions) to favor a strategy of short-term revenue maximization and to have relatively low political/social tolerance for the pain caused by low oil revenues. Countries with small populations and large oil reserves like Kuwait, the United Arab Emirates (UAE), and Saudi Arabia, on the other hand, have tended (also with exceptions) to favor a strategy of long-term revenue maximization and generally have been in stronger positions to weather price declines. Taking into consideration these three factors—producers' interests, consumers' interests, and the diversity within OPEC—the analysis proceeds with a close examination of the recent volatility of oil prices.

The relative stability of oil prices was shattered in 1997 when oil prices plunged to their lowest level in a decade. In addition to the warm winter in the northern hemisphere, the so-called El Nino effect, this huge drop can be explained by three developments, two on the supply side and the third one on the demand side. First, in November 1997 OPEC ministers met in Jakarta, Indonesia, and approved their first increase in the overall production quota since 1993. OPEC's share of world oil supply had remained constant between 1993 and 1996 as non-OPEC supply had increased dramatically. In 1997, OPEC production rose by about 1.6 million barrels per day (b/d), thus increasing its share of the world's oil supply by one percentage point.[23] Saudi officials believed that global demand would rise by 2 million b/d in 1998 and that non-OPEC output would only increase by 1 million b/d.[24] Thus, the decision to raise production can be seen as an attempt by Saudi Arabia to meet growing global demand.

Second, this increase in OPEC production coincided with a change in the Iraqi oil policy. Since the Iraqi invasion of Kuwait and the subsequent Gulf war, Baghdad was banned from exporting its oil. In 1996, however, the United Nations Security Council passed resolution 986 (known as oil for food), which allowed Iraq to sell a certain amount of its oil. Thus, Iraqi oil production rose from 0.5 million b/d in 1996 to 1.2 million b/d in 1997.[25] These additional supplies of Iraqi oil contributed to the glut of the global market. Third, almost at the same time OPEC decided to raise production, the world oil market was stunned by the Asian financial problems. The Southeast Asian economies' demand for oil has grown by nearly 6 percent annually (in comparison with a worldwide rate of 1.7 percent).[26] Before 1997, Asia's economic expansion was the engine driving higher global oil demand. Saudi Arabia's difficulties were strongly compounded by the economic crisis in Asia because Asia accounts for around 60 percent of Saudi oil sales.

There is no doubt that this sharp decline of oil prices was seen as a blessing in the United States and other consuming countries. Low oil prices help to maintain and fuel economic growth. Still, this excessive drop had a devastating impact on the domestic oil industry in the United States. From late 1997 to early 1999, approximately 51,000 jobs were lost.[27] Furthermore, lower prices led to a drop of almost 60 percent in rigs drilling for oil, which was responsible for a decline in the number of new and producing oil wells and resulted in a 7 percent reduction in U.S. proven reserves—the largest percentage decline in more than fifty years.[28] These negative impacts of low crude prices illustrate that the United States, the world's largest oil consumer and importer, has an interest in keeping crude prices at a "reasonable level," not too low. Indeed, when prices fell sharply in 1986, Vice President George Bush was dispatched to Riyadh to convince the Saudi leaders to change their oil policy and stop glutting the market. Shortly after Bush's visit, the Saudi policy was reversed and prices recovered. On the other side, low oil prices to some undefined point can be helpful to Saudi Arabia for several reasons. These include deterring development of alternative energy sources, maintaining Saudi market share against its main competitors, and deterring investments in non-OPEC producing regions.

Still, these extremely low prices substantially reduced state revenues and were behind a massive cut in public spending with potentials for economic and political instability in the long run. In response, the kingdom, in consultation with other OPEC members and non-OPEC producers (Mexico, Norway, Russia, and Oman), took several initiatives to raise prices. In March 1998, the oil ministers of Saudi Arabia, Venezuela, and Mexico met in Riyadh and announced a commitment from most of the OPEC members to boost oil prices by removing up to 1.6 million b/d from the market. Another cut was announced in June 1998. Still, these two attempts failed to impress the market, and prices continued to drop. Finally, in March 1999, oil ministers from Saudi Arabia, Iran, and Venezuela reached an agreement, endorsed by other OPEC members, to further cut production by 1.7 million b/d. In a separate accord, the non-OPEC producers—Mexico, Norway, Russia, and Oman—made a commitment to cut 388,000 b/d.[29] The agreement was effective from April 1, 1999, until March 31, 2000. Global oil markets judged these agreements to be credible, and prices skyrocketed. Several factors contributed to this dramatic swing in oil prices:

(1) The price spike in 1999 can be seen as a direct result of the preceding price collapse. The price collapse of 1998 led to a large number of well closures as well as a reduction in oil exploration and production in non-OPEC countries, including the United States, and it also tended to stimulate oil demand.
(2) The size of the cuts—1.7 million b/d—in addition to over 2.5 million b/d in two output cutbacks agreed to earlier.
(3) Strong world oil demand, including the rebounding Asian economies and the surging U.S. economy.
(4) The decision to curtail output was taken at a time when the supply/demand equation was roughly in balance and there was no further buildup of inventories.
(5) High levels of OPEC compliance to its quota agreement.
(6) The election of President Hugo Chavez in Venezuela in late 1998 brought the country more in line with other OPEC members and in favor of high prices.
(7) Saudi Arabia agreed for the first time since 1993 to set its production well below 8 million b/d, a level that was considered for the

previous six years as immutable irrespective of the supply/demand balance on the market.
(8) The disagreement between Saudi Arabia and Iran about the Iranian quota and actual production levels, which had paralyzed OPEC for several months, was removed by negotiations between the two countries.

The rise of oil prices has had a mixed impact on the United States and Saudi Arabia. As discussed earlier, high oil prices are seen as a drag on the economies of consuming countries. The OECD's rule of thumb is that a $10 rise in the price of a barrel of oil over one year would increase the inflation rate in industrial countries by about half a percentage point and trim a quarter of a percentage point off economic growth.[30] Officials in Washington have taken several preemptive steps to neutralize the economic impact of high prices for oil and other energy resources. Still, high prices have already contributed to a rise in the U.S. trade deficit. Net U.S. oil imports are 4 billion barrels a year, which means that each dollar increase in the price of imported crude oil boosts U.S. expenditures by about $4 billion.[31] Thus, the country's oil import costs rose from $50 billion in 1998 to $67 billion in 1999 and up sharply to $119 in 2000.[32]

On the positive side, once oil producers have replenished their depleted treasury coffers, they are able to buy more of what the industrial world produces, thus keeping the economies of the industrial countries moving ahead. In other words, net oil importing nations earn back some (or much) of the "petrodollars" they originally spend on oil purchases. Another important effect of higher crude oil prices is to bring into production those oil wells and oil fields that were uneconomic at lower prices. This will lead to an increase in output, particularly in high-cost production areas like the United States. In addition, historically, high oil prices are associated with conservation. High prices encourage consumers to use less oil than they usually do.

Finally, it is important to point out that high prices are not in the best interest of major oil producers such as Saudi Arabia. One of the most salient disadvantages could be the development of alternative or competing energy sources, which could undermine the importance of petroleum. In this respect, it is known that when petroleum loses its competitive edge, it will be difficult to recover it even if prices subsequently decline. Another disadvantage of high prices is the exploration and development of oil in high-cost areas in non-OPEC countries, which in turn leads to an increase in supply and exerts downward pressures on prices, which could cause them to collapse. Third, as Ali al-Naimi emphasizes, "We live in one world whose geographical and economic blocks are interrelated and affect each other. Excessively high prices could have a negative impact on the world economy, which in turn weakens demand in the long run and causes producing countries to lose their credibility."[33]

In an attempt to stabilize oil markets and contain price spikes, two important steps were taken, one by the United States and the other by OPEC. In November 1999, Senator Susan Collins (R-Maine) and Charles Schumer (D-N.Y.) introduced Bill 1951, called the Oil Price Protection Act, which would expand the mandate of the Strategic Petroleum Reserve (SPR) to include countering price rises. The proposed legislation would be activated when the price of U.S. crude oil remains above $25 per barrel for two weeks and would require the president to explain what is causing high prices.[34] In September 2000, President Clinton authorized the release of 30 million barrels of oil from the SPR.

A few weeks later, an agreement was announced between the Energy Department and eleven oil companies. Instead of selling oil to the companies for cash, the two sides agreed to "swap." Under a system of swaps, a company would take oil out of the reserve and sell it at current prices. The extra oil on the market would help drive prices down. Later, the company would return a somewhat larger amount of oil to the reserve. The incentive for the companies is that they would earn a profit from selling the reserve oil at today's high prices and replacing it later with cheaper oil bought when prices almost certainly would have fallen. The companies agreed to return 31.56 million barrels to the SPR between August and November 2001 for the 30 million barrels they took in October 2000.[35] Critics of the decision to release oil from the SPR argued that the reserve was created after the Arab oil embargo in 1973–74 to protect against serious disruptions in oil supply, not to soften the blow of lagging inventories or high prices.[36]

The other important move to stabilize global oil markets was taken in March 2000 when OPEC adopted an informal price band mechanism (formally ratified in January 2001) whereby OPEC basket prices higher than $28 per barrel or lower than $22 per barrel would trigger automatic production adjustments. Prices sustained above the target range for twenty trading days are to result in an automatic production increase of 500,000 b/d, while prices below the target range for ten trading days are to result in cuts of 500,000 b/d. OPEC members decided that supply adjustments are not automatic but require approval by an OPEC conference. The price band mechanism was activated in October 2000 to increase aggregate OPEC production quotas by 500,000 b/d.

Together these steps, in conjunction with global economic slowdown and a recession in the United States, succeeded in reducing oil prices by mid-2001. The September 11 terrorist attacks, with their massive negative impact on the aviation industry and the American economy, have dealt a heavy blow to oil demand and prices. Shortly after the terrorist attacks, oil prices went below the $22 mark. This prompted OPEC members to adopt a flexible approach in their demand for higher prices. Indeed, the price band mechanism was suspended in favor of a strategy to counteract the world economic slowdown and the resulting slump in global oil demand.

An Outlook

Several conclusions can be drawn from the preceding analysis of the recent fluctuations of oil prices. First, when the first oil price shock occurred in the early 1970s, there were few major producers. Political and economic instability in the Middle East drove many international oil companies away from the region and caused them to invest in new producing regions. As a result, there are now many oil producers in the global market. Despite the entry of new producers, the bulk of incremental world demand will be met by those countries with the highest reserves (Saudi Arabia and other Persian Gulf producers). As such, Riyadh will continue to have a significant bearing on future capacity additions and management of the world's petroleum supplies and prices.

Second, for decades Saudi Arabia has pursued a policy of price restraint within OPEC. This strategy serves the kingdom's national interest. Keeping world economies dependent on oil supplies at reasonable prices will provide Riyadh with steady revenues and ensure continuity and stability for its economic and political development.

Third, the enormous swings in crude prices since the late 1990s have affected different parts of the world differently. But it has been good for no one. Both extreme low prices and extreme high prices have harmful impacts on oil producers and consumers. The challenge is managing a price that is neither too low nor too high and creating and maintaining a consensus around this price.

Fourth, uncertainty will continue to characterize oil prices in the foreseeable future. Several factors will feed off this uncertainty:

(1) An economic slowdown or a recession in a major consuming economy will likely add to instability in oil prices. The Asian financial crisis of 1997 and the recession in the United States in 2001 are good illustrations. Consumers demand security of supplies, but producers need to feel that there is an adequate security

of demand before making substantial long-term investments in production capacity.

(2) With its huge reserves, Iraq has been a major player in the global oil market for several decades. The continuous political crisis since the Iraqi invasion of Kuwait in August 1990 has caused several interruptions of oil supplies from that country. At an unspecified point in the future, all restrictions on Baghdad will be lifted and the country will resume its role as a major oil producer and exporter. When and how this might happen is unpredictable. Furthermore, speculations on the Iraqi strategy when restrictions are lifted contribute to the uncertainty of energy markets.

(3) Estimates of oil inventories indicate that world oil supply/demand balances do not fully reflect the state of the world oil market. This implies that the stock data are incorrect. This situation has become known as the "missing barrels" problem.[37] The figures vary substantially, both over time and between sources, and there remains a strong disagreement among analysts over the nature of the discrepancy, that is, whether it represents erroneous reporting of production or consumption or actual oil that has not been measured.[38]

(4) OPEC's track record since the inception of quotas in April 1982 indicates how difficult it has been for OPEC to maintain quota discipline. On several occasions, OPEC adjusted quotas to bring them into alignment with OPEC production, rather than the other way around. Quota discipline would bring a sense of predictability to the global oil market.

(5) Finally, one of the major changes in the oil industry has been the gradual erosion of surplus capacity. The growing world demand and the low level of investments in exploration and development resulted in a significant decrease in spare capacity. This lack of capacity to meet growth in demand or to buffer economies against disruptions increases market vulnerability and price volatility. Most of this surplus capacity is concentrated in Saudi Arabia. In order to maintain and develop this significant capacity, the kingdom needs to keep updating and modernizing its oil infrastructure. This has opened the door for potential return of international oil companies to the Saudi energy sector.

The Return of Foreign Investment to the Saudi Energy Sector

In the 1970s, Saudi Arabia gradually and amicably nationalized foreign oil operations run by Chevron, Mobil, Texaco, and Exxon. This step was similar to, but friendlier than, actions taken by many other oil producers. Indeed, during the 1970s virtually all of the oil resources outside of North America passed from international petroleum companies to the governments of oil producers. Each government created its own national oil and gas companies. Since nationalization, Saudi Aramco, the state oil company, has produced and developed most of the oil. Foreign companies were only allowed to participate in downstream operations such as refining and were awarded minor concessions in the Neutral Zone (shared by Kuwait and Saudi Arabia).

The kingdom's adherence to this state-control policy, however, came into question in September 1998 when Crown Prince Abdullah Ibn Abd al-Aziz met in Washington with senior executives from seven American oil companies.[39] He told the American executives that U.S. oil companies had long been the "bedrock" of the U.S.-Saudi relationship and that the Saudi government wanted to join them in a new, strategic energy partnership.[40] To understand this change in Saudi Arabia's energy policy and the American response, the following questions will be addressed: What were the Saudi motives to invite international oil companies (IOCs) back to their energy sector? What were the main characteristics of the so-called Gas Initiative? And what were the main prospects for foreign investment in the Saudi energy sector?

Motives for Inviting Foreign Investment

To be sure, most crude oil producers have switched away from an exclusively state-controlled energy sector to a gradual opening to foreign investment. So the Saudi move should be seen as part of a large trend in the oil and gas industry. Major producers such as Algeria, Indonesia, Iran, and Venezuela have already invited IOCs back to their oil and gas industry. Kuwait, Libya, and Iraq are considering such a move. Theoretically, there are several reasons why governments are reopening themselves to foreign investment: shortages of capital, modern and sophisticated technology, and human resources.[41]

First, the need for capital stems from many factors, the most significant of which is the stable oil prices at a low level from 1987 to 1997, which were followed by a dramatic decline until early 1999. As a result, many producing governments have found themselves without the necessary financial resources to maintain their current levels of production, let alone increase that capacity.

Second, technological advances have drastically changed the oil and gas industry. Most notably, successful explorations have increased, and the costs of production and development of oil fields have fallen. Technological advances have made it cheaper and easier to find and develop hydrocarbon resources. Multinational oil companies have more access to the state-of-the-art technology in the industry than most national oil companies.

Third, human resources include not only management personnel, geologists, and engineers but also professionals with expertise in applying new technologies and new market techniques, such as e-commerce. Some oil-producing countries simply do not have a large enough pool of these necessary experts.

With regard to Saudi Arabia, most of these reasons have contributed to the change in the kingdom's attitude toward foreign investment. More specifically, some considerations should be taken into account in order to understand the Saudi decision to invite IOCs back to the kingdom's energy sector.

(1) Strategic alliance with the United States. The first signs of the change in Saudi energy policy came when the Crown Prince met with top executives of seven U.S. oil corporations. Since then, the negotiations to carry out this initiative have been between the Saudi authority and mostly American corporations. Furthermore, Saudi Arabia has always sought to maintain a large share of the U.S. oil market.[42] For example, in 2000 Saudi Arabia exported 1.57 million b/d of oil to the United States, ranking second (after Canada and just ahead of Venezuela) as a source of total (crude plus refined products) U.S. oil imports and ranking first for crude only (ahead of Canada and Mexico).[43] In short, it is widely believed that American investment in the kingdom and strong oil interdependence between the two countries would underscore and ensure Washington's commitments to the security of Saudi Arabia.

(2) Regional and international competition. Every year, international oil companies spend billions of dollars in exploration and development operations. Recently, more producing regions compete for this badly needed capital, and the competition is getting more

intense not only between different regions (the Caspian Basin, North Sea, and western Africa) but also within the Persian Gulf region. For instance, Iran has already attracted billions of dollars in its energy sector. Iraq is making plans to follow suit once sanctions are lifted. This puts pressure on the Saudi authorities to open their oil and gas sectors to foreign investment. Saudi leaders are well aware that a barrel that is developed in the kingdom by IOCs is a barrel that is not developed by other competitors.

(3) Generating jobs. One of the most daunting challenges facing Saudi leaders is indigenous unemployment. The population is growing at an estimated 3 percent a year, but the rate of increase of new entrants to the labor force is higher because of the youthful profile of the population. Every year more than 120,000 Saudi males enter the workforce, yet the private sector is only creating enough new jobs outside the oil industry to absorb about one in three applicants.[44] The Saudi government is seeking to increase jobs by promoting growth in the non-oil sector. Foreign investments will help to address the unemployment problem, with each billion dollars invested creating between 10,000 and 16,000 new jobs for the Saudis.[45]

(4) Boosting economic growth. Saudi Arabia needs to increase its natural gas production to boost domestic power generation and to fuel new water desalination schemes as well as to serve as feedstock for additional petrochemical plants. This will require approximately $200 billion of investment over the next twenty years.[46] These projects are planned to be wholly financed by the IOCs, without the kingdom's shouldering any financial responsibility.

Gas Initiative

In 2001 Saudi Arabia's proven gas reserves were estimated at 204.5 trillion cubic feet, ranking fourth in the world after Russia, Iran, and Qatar.[47] These massive resources are largely underdeveloped, and official policy has only recently been put in place to fully utilize them. Before 1970, the kingdom's energy industry was dominated by a single energy source: oil. Although gas associated with crude oil was abundant, it was mostly disposed of by flaring as a useless and worthless by-product of oil production. The changes in global energy markets in the mid-1970s prompted the

Saudi government to modify its policy on natural gas. Riyadh instructed Aramco to implement the Master Gas System (MGS), which began functioning in 1981. The MGS fuels and feeds plants in Yanbu on the Red Sea and Jubail on the Persian Gulf. These two industrial complexes account for about 10 percent of the world petrochemical production.[48] The MGS also provides fuel to power utilities, including electrical and seawater desalination plants.

With the discovery of new gas reserves and competitive prices, demand for gas was stimulated in all its uses and grew at a brisk rate of 11 percent annually between 1984 and 2000.[49] As demand is expected to continue growing, a supply strategy was laid down by the Saudi government. Thus, in 1999 Aramco decided to invest $45 billion over twenty-five years on upstream gas development and processing facilities.[50] However, Saudi Arabia realizes that it cannot accomplish its goals without significant foreign investment capital. Under these circumstances, Riyadh has approached IOCs and started the Gas Initiative, which aims to integrate upstream gas development with downstream petrochemicals and power generation and is seen as the key to Saudi Arabia's entire foreign investment strategy.

After the initial contact between Crown Prince Abdullah and U.S. oil executives in 1998, major IOCs submitted proposals for investment in the kingdom hydrocarbon sector.[51] To facilitate negotiations, King Fahd established the Supreme Council for Petroleum and Mineral Affairs in January 2000 with full authority to make final decisions regarding cooperative schemes with foreign companies. In addition, the Council created a ministerial committee, chaired by Foreign Minister Prince Saud al-Faisal, to negotiate with the companies. The committee stressed that the companies' bids will be assessed on a number of points, the most important being their financial positions, technical ability, and capability to handle the projects as well as the extent to which they will provide job opportunities and training for Saudi nationals.[52]

After lengthy negotiations, the Saudi government announced in 2001 that ExxonMobil would take the lead in the $17 billion scheme to explore for and produce gas from any deposits beneath the southern area of Al-Ghawar field.[53] ExxonMobil was also chosen to lead a similar project estimated at $4 billion in the Red Sea area, while Royal Dutch/Shell was awarded the leadership role in the $4 billion Shaybah development in the Empty Quarter desert.[54] There was little surprise that ExxonMobil was chosen to play a prominent role in the gas upstream opening in the kingdom. Exxon and Mobil were two of the original U.S. parent companies that formed the Arabian American Oil Company (Aramco) in the 1940s.

Foreign Investment in the Saudi Hydrocarbon Sector

In early 1999, oil prices rebounded. This added billions of dollars to the treasuries of Saudi Arabia and other oil producers. In spite of the huge oil revenues in 1999–2000, Riyadh needs international oil companies' capital to achieve its ambitious plans to fully utilize its hydrocarbon resources. But the process of building political consensus and the legal regulatory and financial framework to facilitate and accelerate the opening of the kingdom's energy sector to foreign investment is slow and will take time. Nobody expected a swift change. Despite this slow pace, the Saudi government has already taken steps to establish a favorable environment to attract foreign investment. The return of IOCs to the Saudi hydrocarbon sector is a reality. Admittedly, a lot of work still needs to be done to reform the system. The U.S. government has committed itself to support cooperation between American oil companies and Saudi Arabia.[55] Indeed, the Bush administration considers promoting partnership with and investment in Saudi Arabia an essential element in achieving U.S. energy security.

An important explanation for the slow implementation of Gas Initiative is a certain cooling of enthusiasm on both sides (Saudi Arabia and IOCs). In 1998, when oil prices were falling toward $9 a barrel, Riyadh was keener to see a substantial foreign investment in its energy projects. Crown Prince Abdullah was even ready to tempt the companies with the carrots of involvement in the Saudi oil sector. On the other hand, the companies were enthusiastic about tapping into the kingdom's oil reserves—a quarter of the world total. But with the oil price relative increase since 1999, the Saudi mood has changed. When the IOCs eventually presented their plans with a heavy emphasis on oil, they were met with Saudi counterproposals focusing solely on gas, ranging from upstream operations to petrochemical, power, and desalination plants. Saudi domestic natural gas and electricity markets do not offer an attractive rate of return for foreign investors. More accurately, the rate of return is much higher in the oil sector than in the gas sector. However, IOCs accepted the Saudi restrictions on investing in oil exploration and development operations and welcomed the opportunity to focus solely on gas production because they see it as their ticket into the kingdom's hydrocarbon industry. In other words, international companies want to establish a presence in the Saudi energy industry, hoping that one day the oil sector will open and they will be in a more competitive position than those companies that never participated in gas projects.

Saudi Arabia's refusal to open the door for foreign investment in its oil exploration and development operations is quite understandable, given the fact that the kingdom already has a substantial spare crude oil production capacity lying unused. Indeed, Saudi officials have confirmed that in philosophy and principle, the kingdom is not against foreign investment in crude oil production and exploration. But the question is whether foreign company participation would serve the Saudi national interests. There is not much public enthusiasm for finding new oil and increasing the country's production and capacity. Riyadh already has plenty of reserves and idle capacity.[56]

Thus, for the foreseeable future, it is likely that foreign companies' participation will continue to be restricted to the natural gas sector. Still, it is worth repeating that Saudi officials have not ruled out an opening of the oil sector in the future. If such an opening takes place, it will have a tremendous impact on other current and potential producers. With its massive resources and low-cost production, Saudi Arabia is considered the grand prize in the oil industry. A decision to open the Saudi oil sector to foreign investment would erase the interests that foreign companies have in further investments in other producing regions.

Conclusions

The oil industry has changed drastically in the past three decades. There are so many new producers and consumers, a variety of price benchmarks, and a new framework for the relations between international oil companies and crude producers. Two characteristics of the current oil industry need to be highlighted. First, as in the 1970s, today's oil markets are not driven solely by economic forces; they remain highly politicized. Policy and international relations should be taken into consideration. The issue is not the availability of resources; rather, it is creating the geopolitical conditions under which these resources can be fully utilized. The role of producing governments, including Saudi Arabia, in taking over exploration and development operations in the 1970s and the slow reopening to foreign investment illustrate that political factors and economic forces determine oil output. Similarly, the fluctuation in oil prices is less an outcome of a balance between supply and demand and more a reflection of political developments in Riyadh, Washington, and other places. For the foreseeable future, as long as oil is seen as a strategic commodity, policy, not only economics, will continue to shape the oil industry.

The second characteristic of the oil industry concerns the relation between major producers and consumers and the perception of energy security. For a long time, energy security was narrowly defined to mean reducing dependence on imported oil. The relations between producers and consumers were presented in zero-sum terms: The interests of one side could be achieved at the expense of the other side with few, if any, mutual benefits. In today's market, energy security is increasingly defined as a shared issue between producers and consumers. Both will benefit from ensuring that the global energy infrastructure is flexible enough to meet global demand. Consumer anxiety over the availability of supplies is correlated with the producers' concern for secure markets for their crude. In short, the interests of producers and consumers are not mutually exclusive. Both sides aspire to achieve secure supplies and stable prices. Two recent cases can demonstrate this common ground: Kyoto Protocol and the International Energy Forum.

Shortly after taking office in January 2001, George W. Bush announced that his administration would not seek congressional ratification of the Kyoto Protocol. Riyadh supported Bush's decision. Saudi officials estimated that if Kyoto emission cuts were implemented, the national wealth of exporting countries would be reduced by $44 billion by 2010.[57] Saudi Arabia alone would bear $19 billion of this reduction.[58]

The International Energy Forum is an informal gathering of energy ministers from producing and consuming countries with the aim of building confidence, exchanging information, and developing a better understanding of the underlying energy issues affecting the world.[59] It started in 1991 in Paris, and its seventh meeting was held in Riyadh in November 2000. Saudi Arabia proposed the creation of a permanent secretariat for the International Energy Forum based in Riyadh.

This increasing cooperation and growing partnership between Washington and Riyadh to ensure stability in global energy markets will pay particular attention to some of the issues that might threaten the flow of oil supplies from the kingdom. These include security, economic reform, the Arab-Israeli conflict, and militant Islam.

Security

The massive hydrocarbon resources that Saudi Arabia holds and the kingdom's leading role as a moderate force in stabilizing global energy markets leave no doubt regarding U.S. commitments to defend Saudi Arabia. As early as 1943, President Franklin D. Roosevelt stated, "The defense of Saudi Arabia is vital to the defense of the United States."[60] Since then, international and regional developments have confirmed how seriously Washington has carried out this commitment. For instance, in the early 1960s when the civil war in Yemen threatened to spread across the border, the United States sent jet fighters and other military equipment to defend the kingdom. When the Soviet Union invaded and occupied Afghanistan, representing a potential threat to the Persian Gulf region, President Jimmy Carter created the Rapid Deployment Force for possible emergency action. Saudi Arabia, however, objected to granting the United States any air, military, or naval bases. Instead, the Saudis preferred to keep the Rapid Deployment Force "over the horizon."[61] The Saudi leaders did not want to be seen as working too closely with the United States. The Iraqi invasion and occupation of Kuwait and the direct threat it posed to the kingdom changed the dynamics of the Saudi national security. Despite some domestic opposition, the Saudi government officially invited American troops to participate in defending the kingdom and the region. Since then, an unspecified and low-profile U.S. military presence has been established in Saudi Arabia and other Gulf states. In early 2002, it was reported that Saudi Arabia might ask the United States to pull forces out of the kingdom. Some Saudi officials saw the American military presence as a "political liability." Similarly, some senators demanded that Washington consider moving its forces to other countries in the region.[62] Despite these reports, it is hard to imagine any scenario under which American troops would completely withdraw from the Gulf region. Indeed, it is certain that Washington will continue to be the guarantor of military security in the Gulf region and will not hesitate to use all means to protect the kingdom from any external threat. The challenge of internal threats, however, may be harder to manage.

Economic Reform

The Saudi economic system is characterized by heavy dependence on oil, excessive governmental regulations, and efforts to introduce economic reform. First, with oil revenues making up 90 percent of total Saudi export earnings, 70 percent of state revenues, and 40 percent of the country's gross domestic product, Saudi Arabia's economy remains heavily dependent on oil. A major challenge for the Saudi government is to use its huge hydrocarbon resources as a springboard for the creation of wealth by different sectors and sources. Second, the Saudi economy is still dominated by large public corporations such as Saudi Aramco and the Saudi

Arabian Basic Industries Corporation. The kingdom has moved slowly and cautiously toward government subsidy cuts, tax increases, or financial sector reforms. Third, Crown Prince Abdullah and other top Saudi officials have repeatedly confirmed that they see privatization as a "strategic choice" for the kingdom. Several steps had been taken to facilitate and accelerate economic reform including the creation in 1999 of the Supreme Economic Council, charged with boosting investment and creating jobs for Saudi nationals. In 2000, a new foreign investment law was issued, and the Saudi Arabian General Investment Authority was established. It aims at simplifying investment procedures through the creation of a network of one-stop shops, and it gives foreign investors the same incentives and privileges that once were available only to Saudis.[63] The law allows foreign companies to wholly own and operate their own projects.[64] In addition, it reduces the maximum company tax from 42 to 15 percent.[65] In short, the Saudi government is working diligently to establish a favorable environment to attract foreign investment. These steps are in line with larger efforts to join the World Trade Organization (WTO). Riyadh had hoped to be admitted to the WTO by the end of 2000, but the admission was delayed by a variety of issues, including the degree to which Saudi Arabia is willing to increase market access to its banking, finance, and upstream oil sectors. Ultimately, membership in the WTO will result in significant changes toward economic liberalization in the Saudi economy.

The Arab-Israeli Conflict

Nowhere has the gap between the United States and Saudi Arabia been greater than on the Arab-Israeli conflict. Several American administrations have tried to keep energy issues on a separate track from the Arab-Israeli conflict/peace process. Washington has adhered to the notion that the main reasons for instability in the Middle East were Soviet policy (until the collapse of the Soviet Union in 1991) and the lack of genuine efforts to reform the economic and political systems. On the other hand, with their deeply ingrained sense of themselves as the custodians of Arabism and Islam, the Saudis have always retained a sense of kinship with and obligation toward the Palestinians. In spite of sporadic crises and occasional disagreements between the Palestinians and the Saudis, officials in Riyadh have always supported the Palestinian efforts in their conflict with the Israelis. This Saudi stand is also based on internal considerations. Part of the legitimacy of the Saudi royal family and government is based on their adherence to Islam. A large segment of the Saudi population sees the Arab-Israeli conflict in religious terms. Thus, Saudi officials have stressed to their American counterparts that Israel is the main reason for instability in the Middle East.

Not surprisingly, a major crisis in U.S.-Saudi relations was in response to the Yom Kippur War of 1973 between the Arabs and Israel. In one of the lowest points in the relations between the two countries, Saudi Arabia imposed an oil embargo on the United States. Another low point occurred in late 1970s when Egyptian president Anwar Sadat visited Israel and the United States brokered the Camp David Agreement, which was followed by the signing of the first peace treaty between an Arab country (Egypt) and Israel. The Saudi response was to side with radical Arab states in their attempt to isolate and punish Egypt. In 1981, Crown Prince Fahd offered an eight-point peace plan, which called for the Israeli withdrawal from the territories occupied during the 1967 war in exchange for an Arab recognition of the state of Israel. Both the Arabs and the Israelis rejected the Fahd Plan. In the 1990s, Jordan signed a peace treaty with Israel, and two Arab Gulf states (Oman and Qatar) established trade and commercial relations with Israel. There have not been similar movements by the Saudis.

Developments in the Arab-Israeli conflict, including the collapse of the peace process and the beginning of the al-Aqsa intifada (uprising), have further complicated relations between Washington and Riyadh. With nightly news reports of continued clashes between the Israeli army and the Palestinians displayed on Saudi television, anti-Israel and anti-American feelings have risen sharply. Some Saudis believe that the United States has given the Israeli government tacit support for its crackdown on Palestinian violence. They accuse Washington of employing a double standard in its insistence on enforcing the United Nations resolutions against Iraq while not insisting with equal fervor that UN resolutions concerning the Palestinians be fully implemented.

This disagreement between Washington and Riyadh over the Arab-Israeli conflict/peace process is not likely to result in another oil embargo or to seriously threaten oil supplies from the kingdom. Rather, Saudi officials will keep reminding their U.S. counterparts that Washington should take some account of Arab and Muslim sensibilities in formulating its Middle Eastern policy.

Militant Islam

Shortly after the September 11 terrorist attacks, President Bush repeatedly stated that the world has been divided into two camps, good and evil. Each country has to define where it stands: "Either you are with us, or you are with the terrorists." The official Saudi position has left no doubt that

Riyadh strongly condemns the attacks and supports the war on terrorism. Officials in the Bush administration have repeatedly expressed their satisfaction with Saudi cooperation in the war against terror. However, Congress and several American news organizations have criticized the kingdom's domestic and foreign policies because more than a dozen of the hijackers were Saudi citizens and their al-Qaeda operation was partly financed with private Saudi money. Furthermore, FBI agents routinely expressed their frustration at the lack of Saudi cooperation during the investigation in previous attacks (e.g., the 1996 Khobar Towers housing complex in which nineteen American servicemen were killed).

To understand where the Saudi Arabian government stands on the question of militant Islam, consider this: First, the kingdom's political and social system is one of the most conservative in the Islamic world. The Saudi state has been built on an alliance between a religious movement known as Wahhabism and the Saudi royal family. The legitimacy of the latter has been intimately linked with the main tenets of the former. Thus, the Ulama (religious scholars) have always had some leverage in formulating the Saudi domestic and foreign policies. Their powerful constituency cannot be ignored. Some were reluctant to endorse the war against Afghanistan, a fellow Muslim country. As the Grand Imam (religious leader) stated, "This issue [terrorism] calls for new policies, not new wars."[66] Second, it is important to remember that one of Osama bin Laden's chief goals was to topple the Saudi monarchy, which he considered corrupt and un-Islamic because it is allied with the United States and has allowed American troops to be stationed there since the 1991 Gulf war. In other words, militant Islam is as much a threat to the Saudi government as it is to the United States. The two sides stand together against militant Islam. In the future, the Saudi government is likely to continue walking a tightrope in order not to antagonize its domestic constituency while maintaining strong ties with its major international ally, the United States.

Security, economic reform, the Arab-Israeli conflict, and militant Islam will probably continue to affect the friendship between the two nations. For the foreseeable future, officials on both sides are likely to continue working together on security and energy issues. The world's largest oil exporter and the world's largest oil importer need one another.

4

The United States and Iraq

Continuity and Change

For much of the past fifty years, suspicion, tension, and hostility have characterized relations between the United States and Iraq. After the defeat of the Ottoman Empire in the First World War, Iraq was created as a nation-state. It became formally independent in 1932, but the British influence remained dominant until the overthrow of the monarchy in the nationalist revolution of July 1958. The government in Baghdad was a strong supporter of the Western powers, particularly Britain. In 1951, Iraq successfully approached the United States for economic and military assistance. Four years later, Iraq, Iran, Turkey, Pakistan, and Britain signed the Baghdad Pact. The alliance formed a defensive cordon along the southern fringe of the Soviet Union. General Abd al-Karim Qassim's rise to power in 1958 presented a dramatic shift in Baghdad's foreign policy orientation. Shortly after the establishment of the republic, Iraq withdrew from the pact, and relations with the United States became tense. Diplomatic ties were finally severed in 1967 with the outbreak of the Arab-Israeli war in which Washington backed Jerusalem.

The Iraqi political scene then witnessed a domestic power struggle between nationalist and leftist groups, with each expressing deep distrust and hostility toward the United States. In addition, Baghdad adopted an uncompromising attitude against the Arab-Israeli peace process led by Egypt and Syria. Furthermore, Iraqi relations with the traditional Arab monarchies in the Gulf and with the shah of Iran were characterized by mutual suspicion. Finally, Baghdad consolidated its anti-Western policy by signing a friendship treaty with the Soviet Union in 1972.

The rise of Ayatollah Khomeini in Iran in 1979 altered the political dynamics in the Gulf region. Suddenly Baghdad was perceived as the bulwark against the messianic Islam of the mullahs. The growing hostility

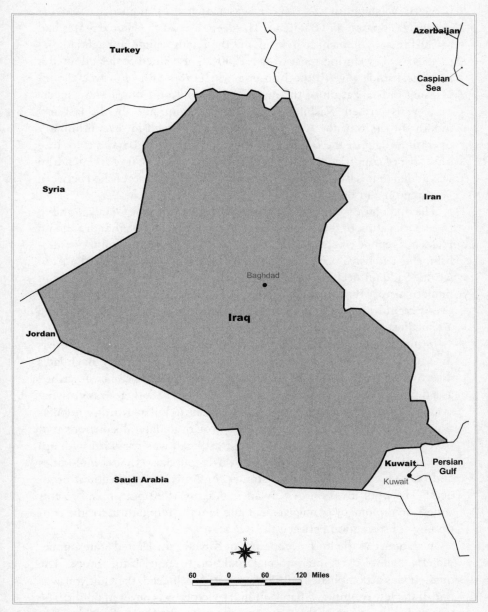

Map 4. Iraq

between Washington and Tehran was matched by rapprochement between Washington and Baghdad. The deep distrust the United States had toward the government in Tehran and the Baath regime in Baghdad dictated its policy during most of the 1980s. The United States adopted a neutral attitude toward the Iran-Iraq war (1980–88). However, the increasing radicalization of the mullahs and the Iran-Contra affair in the mid-1980s pushed Washington toward the Iraqi side. In the last few months of the war, the American navy was involved in several military operations against the Iranians. Meanwhile, the United States took Iraq off its list of countries sponsoring terrorism in 1982. Two years later, Iran was added to this list, and diplomatic relations were reestablished between Washington and Baghdad.

The outcome of the Iran-Iraq war paved the way to the Gulf war and to the deterioration of relations between Washington and Baghdad. Saddam Hussein's regime emerged from the war with an enormous military establishment, extensive debts, and an unchecked ambition to lead the Arab world. Saddam declared that the war against Iran was the first time in modern history that an Arab country had defeated a non-Arab power. The overly confident Iraqi leader started taking a strong position against Israel and against American influence in the Middle East.

From the end of the Iran-Iraq war until the beginning of the Gulf crisis in August 1990, Washington tried unsuccessfully to modify Baghdad's assertive policy. During these two years, American policy can be described as a diplomacy of tolerance and restraint. In late 1988, there was growing evidence that Iraq was using chemical weapons to kill its Kurdish population. In response, the U.S. Senate passed tough unilateral sanctions that have never been implemented.[1] By contrast, Iraq was provided with approximately $1 billion of credit per year to purchase U.S. commodities and became the ninth largest customer for American agricultural products.[2] The Iraqi invasion of Kuwait in August 1990 put an end to this American diplomacy of engagement and laid the foundation for the continuing confrontation between the two states.

In response to the Iraqi occupation of Kuwait, the United States immediately assembled an international coalition to expel Iraqi troops. The mission was accomplished with little resistance. Indeed, the Gulf war was one of the few examples where a military victory was never in doubt. The war ended, however, without serious efforts to overthrow Saddam Hussein. Many in Washington and elsewhere mistakenly thought that the Iraqi leader's days were numbered. Furthermore, the United Nations mandate was to liberate Kuwait, not to seek any political change in Baghdad.

Finally, President George Bush wanted to maintain an "exit strategy" for the United States from the Gulf crisis and to keep the international coalition intact without any defection from Arab states. A few years later, Bush admitted that he had underestimated the political staying power of Saddam after the Gulf war and "regretted that the allies did not do more to undercut his authority."[3]

Regardless of Washington's reasons not to pursue efforts to overthrow the Iraqi leader, the Gulf war ended with Saddam at the helm in Baghdad. Against all odds, Saddam has been able to recover and reestablish control over most of Iraq. The regime in Baghdad has continued to pose a serious threat to U.S. interests in the Gulf region. Washington's hope of a change in the Iraqi leadership and of a friendlier regime in Baghdad has not been realized.

The continuous confrontation with Iraq since 1990 (indeed, the troubled relations since 1958) underscores the contradiction between two fundamental goals of U.S. policy in the Persian Gulf. On one side, allowing Iraqi oil shipments to global markets would contribute to stability of energy supplies and prices. Today Iraq contains 112 billion barrels of proven oil reserves, the second largest in the world (behind Saudi Arabia) along with roughly 215 billion barrels of probable and possible resources.[4] Iraq is a crucial player in the regional and global oil industry.

On the other side, for historical and geographical reasons, Baghdad has always advocated changes in the status quo in the Gulf region. Iraq's territorial disputes with Iran and Kuwait have contributed to regional instability. In the past two decades, Iraq had been to war twice, partly to resolve such conflicts. In 1980 Saddam Hussein wanted to exclude the Iranians and to have unilateral control over the Shatt al-Arab waterway. After eight years of fighting, he came very close to achieving this goal. However, to avoid war on two fronts—the Iranian and the Kuwaiti—during the Gulf crisis of 1990–91, Saddam decided to share the waterway again with Iran. Baghdad had similar disappointing experiences in revising its borders with Kuwait. Since the late 1930s, Iraq has launched a number of campaigns to absorb Kuwait. In the process, Baghdad has never agreed to finalize any border demarcation with the emirate. During the 1980s, Kuwait supported Iraq in its war against Iran. The Iraqi troops were allowed to use the two strategic islands of Bubiyan and Warba. Kuwait refused the Iraqi request to lease the two islands. After the Gulf war, in November 1994, the defeated Iraq had no choice but to recognize Kuwait within the borders defined by the United Nations. Regional borders were redrawn in favor of Iran and Kuwait. This suggests that once the opportunity comes, the Iraqi leadership will not hesitate to challenge the status quo and again threaten regional security. Little wonder the United States and other countries perceive Iraq as a threat to the security in the Persian Gulf that needs to be contained and neutralized. The challenge is how to guarantee the steady flow of Iraqi oil to the international markets while preventing the Iraqi leadership from using oil revenues to build a massive military capability including weapons of mass destruction.

This chapter examines American efforts to pursue these two goals—an interest in Iraq's oil resources and an attempt to neutralize Baghdad's threat to regional security—as well as the Iraqi government's policy of maximizing oil revenues and asserting what they perceive as legitimate sovereign rights. The focus will be on the period since the Iraqi invasion of Kuwait in 1990. After more than ten years of containment, the United States still has no clear strategy for dealing with Iraq. The intense debate in Washington in the aftermath of September 11 underscores a twofold conclusion. Most U.S. policymakers have embraced the idea of a regime change in Iraq. The disagreement, however, concerns how to achieve this goal and whether it can be achieved at an acceptable political, military, and diplomatic cost.

The Iraqi Oil Industry

Despite the fact that Iraq has massive proven oil reserves and that petroleum was discovered in Iraq earlier than in most other Middle Eastern countries, the country's hydrocarbon resources are largely underexplored and underdeveloped. Oil was first struck in commercial quantities at Naft Khana adjacent to the Iranian frontiers in 1923.[5] Shortly after that, the Turkish Petroleum Company (TPC) was confirmed in its concession covering most of Iraq. The shares in this firm were divided between British, Dutch, and French partners. Significantly, no American individuals or companies participated in the TPC.[6] In 1927, a giant oil structure was discovered in Kirkuk. These promising oil resources attracted American firms. Accordingly, in the late 1920s the composition of the TPC was changed. Five American oil companies acquired shares,[7] and the new firm was renamed the Iraqi Petroleum Company (IPC). Thus American oil companies had participated with their European counterparts in developing Iraqi oil resources at an early stage.

The demise of the monarchy in 1958 and the rise to power of nationalistic and leftist regimes had drastically altered the relations between the

Table 4.1. Iraq's oil production, 1970–2000 (thousand barrels per day)

Year	Production	Year	Production	Year	Production
1970	1,549	1981	1,000	1992	425
1971	1,694	1982	1,012	1993	512
1972	1,466	1983	1,005	1994	553
1973	2,018	1984	1,209	1995	560
1974	1,971	1985	1,433	1996	579
1975	2,262	1986	1,690	1997	1,155
1976	2,415	1987	2,079	1998	2,150
1977	2,348	1988	2,685	1999	2,508
1978	2,563	1989	2,897	2000	2,571
1979	3,477	1990	2,040		
1980	2,514	1991	305		

Source: Energy Information Administration, *International Petroleum Monthly* (Washington, D.C.: Government Printing Office, May 2001), 38.

Iraqi government and IPC. An important step in this direction was the issue of Public Law 80 of 1961 under which the Iraqi government seized approximately 99 percent of the concession territory of the IPC and its affiliates. A few years later, a state-owned company was established, the Iraq National Oil Company (INOC). Finally, in 1972 the Iraqi government nationalized the IPC, and by 1975 the holdings of various private companies working in Iraq were completely transformed to the INOC. Since the mid-1970s, the INOC has taken full control over the country's oil industry; international oil companies have only been awarded service contracts.

In the following years, Iraq's oil production reached its peak, but these favorable conditions did not endure. Since 1980, the Iraqi oil industry has been a victim of two wars as well as a prolonged and comprehensive economic sanctions regime. Consequently, the country's level of production has deteriorated, reflecting these unfavorable political and military developments. Since the late 1990s, Iraqi oil production has risen in response to several resolutions passed by the UN Security Council. Table 4.1 illustrates the impact of wars and political problems on Iraqi oil production. The peak of 1979 (3.5 million barrels per day) was reached before the crises of the following two decades. The huge drop in 1981 was in response to the war with Iran in which the two sides attacked each other's oil installations. The substantial increase in production in 1989 coincides with the only year when Iraq was not at war with any of its neighbors.

This relative peace did not last long, however. The Gulf war and economic sanctions took a heavy toll on Baghdad's oil industry. Both upstream and downstream installations were targets for the international alliance's missiles and bombs. Furthermore, since oil was the main source of foreign currency revenue, it became the main focus of the sanctions. Iraqi oil exports were restricted to supplying Jordan with limited quantities. The rise in production in 1997 was the result of UN Security Council Resolution 986 (also known as oil for food). In April 1995, the UN Security Council passed Resolution 986, which allows limited Iraqi oil exports for humanitarian and other purposes. Iraq actually began exporting oil under Resolution 986 in December 1996. Proceeds from the program are used to pay compensation for Gulf war victims, pipeline transit fees for Turkey, funding for UN weapons monitoring activities, costs of operating the Iraqi oil industry, and provision of UN-approved goods to Iraq. With Iraq steadily increasing its oil export revenues, the UN Security Council passed Resolution 1284 in December 1999 to remove any limits on the amount of oil Iraq could export.

These continuous political crises meant that few financial resources were available to renew the country's energy infrastructure. Iraq was heavily underexplored and underdeveloped. Even worse, it is not known how badly the oil wells were damaged by wars and sanctions. A recent study commissioned by the United Nations estimates that "as many as one-fifth of Iraq's oil wells might be irreparably damaged."[8] Indeed, one of the major problems that have been plaguing the oil industry in Iraq since the early 1990s is the embargo placed on the import of necessary equipment and spare parts. In order to address this problem, the UN Security Council adopted Resolution 1175 in June 1998 authorizing, for the first time since the Gulf war, the import of up to $300 million in equipment and spare parts for the oil sector every six months. This was not enough to stop the deterioration in Iraq's oil industry. Accordingly, in January 2000 UN Secretary-General Kofi Annan sent a team of experts to inspect Baghdad's petroleum fields, pipelines, and refineries. The team concluded that the entire program for Iraq's oil industry was "out of step with modern engineering techniques and generally accepted principles of 'value for money' investment."[9] They added, "The ability of the Iraqi oil industry to sustain current reduced production levels will be seriously compromised unless effective action is taken immediately."[10] In response to this report, the UN Security Council adopted Resolution 1293 in March 2000, which raised the cap on imports for the oil sector to $600 million every six months.

Political and military crises have not only restrained Iraq's level of production but also seriously reduced its ability to ship its crude to the international markets. Iraq is almost a landlocked state with a narrow outlet on the Persian Gulf, where it has three tanker terminals: Mina al-Bakr, Khor al-Amaya, and Khor al-Zubair.[11] The infrastructures for these terminals were almost completely destroyed in the war with Iran and the Gulf war. Since then, substantial repairs have been done. Still, these terminals do not have enough facilities for the Iraqi massive oil exports. The country has had to rely on pipelines that traverse other countries. These include Israel/Palestine, Syria, Saudi Arabia, and Turkey.

Shortly after oil was discovered in commercial quantity in Iraq, the IPC built a pipeline to Haifa in British-controlled Palestine and another pipeline to Tripoli in Lebanon with a sideline forking from Homs to Banias in Syria.[12] The establishment of Israel and the resulting Arab boycott caused the shutdown of the Haifa pipeline. The civil war in Lebanon (1975–89) raised security concerns regarding the pipeline to Tripoli, so the pipeline was closed for a number of years. During the Iran-Iraq war, Syria supported Tehran and in 1982 closed the Tripoli/Banias pipeline to deny Baghdad oil exports and revenues. This pipeline remained closed until late 2000. Anticipating troubles in its relations with Syria, Iraq decided to take a protective measure. It built two pipelines to the Mediterranean via the territory of Turkey (Kirkuk-Ceyhan) in the 1970s and 1980s. The pumping stations and other facilities for these pipelines were destroyed during the Gulf war, and since then massive repairs have been done.

In addition to these pipelines to the Mediterranean Sea, Iraq constructed two pipelines (in 1985 and 1990) to the Red Sea through Saudi Arabia. Both were owned entirely by Iraq but were shut down by Saudi decision in August 1990, when Iraq attacked Kuwait. In June 2001, Saudi Arabia decided to expropriate the two pipelines. The reason, according to Riyadh, was "the continued Iraqi threats of aggression in the years after the occupation of Kuwait and the Gulf war."[13] Finally, in order to optimize export capabilities, Iraq constructed a reversible "Strategic Pipeline" in 1975. This pipeline consists of two parallel lines. The North-South system allows for export of northern Kirkuk crude from the Persian Gulf and for southern Rumaila crude to be shipped through Turkey. During the Gulf war, the Strategic Pipeline was disabled.[14]

To sum up, for a long time the Iraqi oil industry has been the victim of political and military conflicts. Due to heavy financial constraints, Baghdad has not been able to develop production capacities to levels compatible with its huge reserves or to construct and upgrade its pipeline systems. The economic sanctions, imposed since 1990, have substantially contributed to the deterioration of the Iraqi oil industry.

Economic Sanctions

On August 6, 1990, only a few days after Iraq invaded Kuwait, the UN Security Council passed Resolution 661. This banned the purchase or transshipment of Iraqi oil and other commodities and the sale or supply of all goods and products to Iraq, with possible humanitarian exceptions for medical supplies and foodstuffs. Initially, the reason for sanctions was to force Iraq to withdraw from Kuwait. These sanctions were observed by almost all members of the international community, particularly in the early 1990s. Shortly after the Gulf war, the rationale for maintaining sanctions was shifted to Iraq's weapons of mass destruction (WMD). UN Security Council Resolution 687, adopted April 3, 1991, delineates terms of cease-fire and conditions for lifting the sanctions. The Resolution required that Iraq remove or destroy all of its nuclear, chemical, and biological weapons, as well as ballistic missiles with a range of more than 150 km. It also created the United Nations Special Commission (UNSCOM) to supervise and document Iraq's disarmament. In other words, Resolution 687 outlined two principal tasks: for Iraq to declare and for UNSCOM to verify and supervise the elimination of its prohibited weapons. Sanctions were the way to convince Iraq to cooperate with inspectors. This was a combined carrot-and-stick approach. Keeping the sanctions was the stick, and the carrot was that if Iraq cooperated with the elimination of its WMD, the Security Council would lift the sanctions.

The United States has viewed sanctions not only as a means to achieve this specific goal but also as a way to intensify economic pressure on the regime in Baghdad by depriving it of badly needed financial resources. Bluntly, Washington has viewed sanctions as part of a broader effort to undermine Saddam Hussein's regime and eventually remove him from power. Thus, the rationale for imposing and maintaining sanctions has evolved from forcing Iraq to withdraw from Kuwait to pressuring Baghdad to cooperate in destroying its WMD and finally as a means to overthrow Saddam Hussein and undermine his regime.

The sanctions regime was supposed to be a temporary measure to achieve a specific goal. When sanctions were imposed in 1990, no one expected them to last as they have. Indeed, there is a growing consensus that the sanctions regime is unprecedented in its comprehensiveness, severity, and length and in the enormous human and economic costs that have been inflicted on Iraq. In an early attempt to alleviate some of these devastating social and economic consequences, the UN Security Council adopted Resolutions 706 and 712 in 1991. These two resolutions proposed that Iraq should be allowed to sell oil worth up to $1.6 billion over six months. This money was supposed to be used for the purchase of food and medicines. Baghdad's response was a strong rejection on the basis that this represented a violation of its sovereignty. Four years later, with the sanctions' impact increasingly felt, Resolution 986, similar to the previous ones, was adopted. In May 1996, Baghdad accepted the deal, which allowed it to sell up to $1 billion worth of oil every three months.[15] About two thirds are spent on food and medicine. The other third is spent on war reparations and UN expenses in Iraq. But the implementation of this resolution was delayed several months because of disagreements between Washington and Baghdad. Finally, in December 1996, Iraqi oil was allowed into the international market. Iraq portrayed the deal as the first step toward lifting the sanctions, whereas the United States insisted that the sanctions would remain until Baghdad complied fully with all UN resolutions. Since then, the program has been renewed several times.

Given the large share of oil revenues in Iraq's national income and as a source of foreign exchange, restricting oil exports has had a devastating impact. A close examination of more than a decade of sanctions should take into consideration the socioeconomic, military, and political consequences. First, sanctions have transformed Iraq from a country of relative affluence to a country of massive poverty.[16] According to the United Nations Development Program, Iraq has the lowest life expectancy and adult literacy rates among the eight Persian Gulf states. Furthermore, Iraq is the only country where the infant mortality rate is on the rise.[17] Two heads of the UN humanitarian program in Iraq—Denis Halliday and Hans Van Sponeck—resigned in protest against the sanctions. Describing the impact of sanctions, Halliday wrote, "Children have been forced to work, to beg and engage in crime and young women have been forced into prostitution by the destruction of their families."[18] It is also important to point out that these socioeconomic hardships are felt the strongest among ordinary citizens with little impact on members of the Iraqi elites. As one observer summed it up, "Economic sanctions and the resulting shortages have created a small group of winners as well as a large number of losers."[19]

Second, until the Gulf war, Iraq was among the world's top arms importers. Putting aside speculations on the quantity and quality of Baghdad's conventional weapons as well as how many WMD it still maintains, it can be argued that the lack of necessary funds has severely hindered Iraq's ability to rebuild and modernize its conventional military and has thwarted its pursuit of nuclear weapons. One can speculate that at best Iraq has made limited progress on its WMD programs, particularly when compared with their rapid development in the 1980s.[20]

Third, Saddam Hussein has succeeded in creating a perception that the deteriorated socioeconomic conditions in Iraq are the result of sanctions, not his failure to comply with the UN resolutions. This wide perception of Iraqi children starving because of a lack of basic food and medicines has contributed to dissatisfaction and opposition to the U.S. policy, particularly in the Arab and Muslim worlds.

In addition to Saddam Hussein's campaign to publicize the suffering of the Iraqi people and to squarely blame the United States for insisting on maintaining the sanctions, two developments have contributed to the weakening of the sanctions regime and strengthened his propaganda against Washington. First, the relative shortage of oil supplies in the global markets in the late 1990s has been considered a twofold blessing for the Iraqis. Low inventories in consuming countries and lack of surplus production capacity in producing countries have made the continuity and noninterruption of oil shipments from Iraq and from other producers crucial for the stability of oil markets and prices. In other words, maintaining a maximum level of production for every producer becomes crucial in a tide oil market. On the other hand, this shortage of supplies is a major reason for high oil prices. The rise in prices (early 1999 to mid-2001) added billions of dollars to the Iraqi coffers. This substantial increase in oil revenues has lured many countries to seek business opportunities in Iraq and gradually facilitated and accelerated Baghdad's efforts to break the international isolation imposed by the United Nations and strongly supported by the United States since 1990.

The other important development that has strengthened Saddam Hussein's propaganda against the United States in recent years is the breakout of intense violence between the Israelis and Palestinians, the so-called Al-Aqsa intifada (uprising). American support for Israel has played into Hussein's hands. With his firm stand against the United States, Hussein is seen as a hero in some Arab and Muslim countries. The Clinton administration sought to separate developments in the Persian Gulf from developments in the Arab-Israeli conflict. Recent developments in both regions demonstrate their inseparability. The rise in tension in the Arab-Israeli conflict would be echoed in the Gulf region.

In light of these developments—the rise in crude prices and intense violence between the Israelis and the Palestinians (as well as growing tension between Israel and Syria)—Baghdad has taken several steps to further loosen the straitjacket that it has been in for more than a decade. These include replacing the American dollar by Europe's joint currency, the euro, for payments for its oil exports, imposing the surcharge, and reopening the Kirkuk-Banias pipeline, which had been closed since 1982.

(1) In late 2000, the Iraqi government decided to ban the use of U.S. dollars in the country. It also decided to transfer its UN-monitored escrow petroleum account from U.S. dollars to euros and to halt all foreign trade transactions in the American currency. The rationale behind this move is to minimize the influence Washington has on Baghdad's international economic and financial transactions. In line with this new policy, the Iraqi government told its oil customers to pay for crude in euros beginning in November 2000. A report by the UN Treasury Department found that the shift would incur high transaction costs.[21] Still, Baghdad insisted on this change, and the UN Sanctions Committee approved the Iraqi request.

(2) In November 2000, Iraq's State Oil Marketing Organization demanded that companies lifting cargoes of Iraqi crude oil begin paying a 50 cent per barrel surcharge directly to the Iraqi government. The purpose of this request is twofold. The surcharge will create revenue streams independent of the UN account and give the regime in Baghdad revenues it can spend without international oversight. Furthermore, these extra payments would further reduce the credibility of the sanction regime. Later, the initial surcharge was reduced after Baghdad found it difficult to collect. Many companies have been reluctant to pay the surcharge even when offered terms that would put the total price per barrel including the surcharge at below-market levels. The reason is that companies do not want to be punished for violating UN sanctions.

(3) Since the late 1990s, relations between Baghdad and Damascus have substantially improved. Top officials from the two countries have exchanged visits and signed cooperation agreements, including a free economic zone accord in January 2001. Syrian exports of crude oil suddenly spiked in late 2000. This pointed to a new

and significant cooperation between Syria and Iraq. Apparently Syria was using Iraqi supplies to meet local demand and exporting oil that it produced domestically. It was reported that the two countries had reopened the Kirkuk-Banias pipeline. According to some analysts, Syria was receiving between 120,000 to 200,000 barrels per day (b/d) in a sharply discounted price. These could be yielding approximately $2.5 million in revenue daily to the Iraqi regime outside the UN-controlled account.[22] In addition to a specific amount of oil Iraq is allowed to export to Jordan, two export points—one in Turkey and the other in the Persian Gulf—have been under UN supervision since the mid-1990s. The Syrian outlet was not licensed by the Sanctions Committee. The Bush administration has encouraged Syria to bring the pipeline into conformity with the sanctions and has offered to allow Damascus to import Iraqi oil as long as the revenue is deposited in a UN account and does not go to the Iraqi leadership.

These three steps—using euros instead of U.S. dollars, adding a surcharge, and reopening the Kirkuk-Banian pipeline—are in line with a strong conviction held by the Iraqi leadership that sanctions would not be removed by a political decision within the UN Security Council. Rather, the Iraqi leaders believe the sanctions regime will erode by itself and disintegrate. Saddam Hussein has repeated and confirmed this conviction in his speeches since the late 1990s.

Outlook for the Iraqi Oil Industry

The increasing dislike of the sanctions, the rise in oil prices in the late 1990s, and the collapse of the peace process between Israel and the Palestinians and Syrians have all helped Iraq to improve relations with regional and international powers. An important step in breaking the international isolation took place in August 2000, when the Iraqi government reopened Saddam International Airport. Since then, flights from Arab, Asian, and European countries have landed there, demonstrating the steady erosion of the sanctions. Under these conditions, three trends can be identified in the direction of the Iraqi oil industry. These are Iraq's emerging role as a "swing" producer, Baghdad's plans to attract foreign investment to its oil industry, and George W. Bush's strategy to deal with the Iraqi threat.

First, for much of the 1990s, one of the energy industry's most crucial questions was when would Iraqi crude return to the market? This question is not valid, partly because so much Iraqi crude is already available in the market. Also, how the Iraqi oil will reach the market (under UN supervision or outside its control) will affect the stability of global oil supplies. Under the UN sanctions, Iraq is supposed to sell its crude only under a program that deposits all proceeds into an escrow account. Despite these arrangements, Iraq has emerged as a significant swing producer, and Saddam Hussein has demonstrated willingness to manipulate oil markets. In 1997 and 1998, rapidly increasing Iraqi oil exports played a significant role in creating a world oil glut and causing a price collapse. Under UN supervision, Iraqi oil is commonly sold to Russian, Chinese, Malaysian, Italian, and French firms. Oil is then resold to interested parties, including American oil companies. Indeed, Iraq has rapidly emerged as an important crude supplier to the United States.

In addition to this oil sold under UN control, Baghdad has expanded its smuggling operations via a number of routes including Turkey, Jordan, Iran, and Dubai. Syria is not the only avenue for smuggled Iraqi oil. Hundreds of thousands of barrels of oil have been smuggled from northern Iraq into Turkey. But because Turkey provides a crucial air base used by American aircraft patrolling over Iraq, U.S. officials are hard-pressed to demand that Ankara crack down on the lucrative trade. As one analyst noted, "Everybody knows that Saddam is cheating like crazy, and nobody really wants to do anything about it because we all have a vital interest in seeing as much oil as possible flow out of Iraq."[23] Furthermore, when sanctions are finally lifted, it is unlikely that Iraq will participate in OPEC production constraints because of the need to rebuild its economy. Instead, most likely Baghdad will raise capacity to maximize its revenues and to reflect its huge proven reserves. In order to achieve this goal, Iraq will need massive foreign investment in its oil industry.

Second, Iraq is likely to attract massive foreign investment in its oil industry once sanctions are lifted. The country's huge reserves are largely underdeveloped. Furthermore, there are no indigenous financial assets to meet the requirements for badly needed investment. Indeed, shortly after the end of the war with Iran, the Iraqi government was hoping that foreign companies would participate in the development of some of its fields. In February 1990, a few months before the invasion of Kuwait, Baghdad specified the terms under which foreign companies could operate: Contractors were to finance the work and to be repaid out of the subsequent production of the fields they develop once they came on stream. The Gulf war brought to a halt these early efforts to liberalize Iraq's hydrocarbon sector. This halt, however, proved to be temporary. The huge damage to the country's economic and social infrastructure caused by the Gulf war and the sanctions suggests that the main way for recovery is to maximize oil revenues.

In an attempt to reward friendly states and punish hostile ones, the Iraqi policy since the late 1990s has been to award gas and oil concessions to companies from countries supporting the easing or lifting of UN sanctions, particularly France and Russia. Iraq owes France some $5 billion for earlier shipments of weapons and supplies.[24] The first major order of Western arms dates from September 1976, when France agreed to supply between 60 and 80 Mirage F1s, followed by an order for 200 French AMX 30 tanks in 1977.[25] Since then, Paris has developed special relations with Baghdad, and France was a major arms supplier during most of the 1980s. Since the Gulf war, France has adopted a softer stance against Iraq in comparison with that of the United States and Britain. Furthermore, Iraq was the birthplace of the giant French oil company Total, where its ancestor, Cie Francaise des Petroles (CFP), began work in 1924, after acquiring the German shares of the Turkish Oil Company. Even after nationalization of the Iraq Petroleum Company in 1972, France offered to buy Iraqi crude at a special price for ten years as an agreement between the Iraqi and French governments.[26] France supports permitting foreign investment in the Iraqi oil industry.

Like France, Russia has pursued a more assertive policy in the Middle East since the second half of the 1990s. Moscow has had close ties with Baghdad since the late 1950s. The two countries formalized their relations by signing a friendship treaty in 1972. The Soviet Union was the main source of armaments to Iraq for most of the two decades preceding the Gulf war. Due in large part to arms sales, the Iraqi debt to Russia is estimated at $10 billion.[27] Despite UN sanctions restricting Iraqi oil sales, a Russian energy company—Zarubezhneft—announced that it was drilling scores of oil wells in Iraq.[28] The drilling, which is concentrated in the Kirkuk field in northern Iraq, is the first by a foreign company since sanctions were imposed. In addition, in September 2001, Baghdad announced that Russian companies had won deals worth $40 billion to develop scores of future oil and infrastructure projects.[29]

This cooperation with French and Russian companies, as well as companies from other countries, is part of an Iraqi strategy to increase its oil production capacity to 6 million b/d, following the lifting of the UN sanctions. According to an Iraqi former oil minister, Fadhil J. Chalabi, $30 billion in upstream investment is needed over a decade for the country to reach the officially targeted production level of 6 million b/d.[30] Thus, since the late 1990s, Baghdad has negotiated and signed several multibillion-dollar deals with foreign oil companies, not only from France and Russia but also from China, Italy, and other countries. Three important characteristics of these deals need to be underscored.

(1) The UN Sanctions Committee does not object to the signing of agreements of intent with Iraq as long as no services are actually provided and no financial transactions take place.
(2) Iraq has become increasingly frustrated by the failure of foreign companies to begin work on the ground and has threatened not to sign deals unless firms agree to start drilling without delay.
(3) In order to address this problem and to lure more foreign investment, the Iraqi Oil Ministry introduced amendments to existing contracts. Among other things, a clause was added referring to "an explicit commitment to achieve target production within a set period." In addition, participation by an Iraqi entity would be 10 percent, versus 25 percent in previous contracts.[31]

The third trend in the outlook for the Iraqi oil industry is the new approach adopted by the Bush administration in 2001. Top officials in the Bush administration particularly criticized two dimensions of the Clinton administration's policy on Iraq—the end of UN arms inspections in 1998 and the gradual erosion of support for sanctions. The fact that several members of the Bush administration are veterans of the Gulf war incited expectations that the new national security team is likely to adopt an aggressive stand in relations with Iraq.[32] This became apparent from several statements that Secretary of State Colin Powell made during his confirmation hearing. Powell described Iraq as a "failed state with a failed leader." He added, "We need to be vigilant, ready to respond to provocations, and utterly steadfast in our policy toward Saddam Hussein. And we need to be supportive of opposition efforts."[33]

Shortly after taking office, the Bush administration proposed a new approach toward Iraq. The core of this new approach is a modification of the sanctions regime. In testimony before the House International Relations Committee, Powell stated that the practical difficulties and political cost of maintaining the current embargo "are beginning to outweigh its usefulness."[34] Accordingly, the Bush administration proposed the so-called smart sanctions. The main characteristics of these smart sanctions are to lift sanctions on civilian goods, retain sanctions on dual-use goods, increase international control on Iraqi oil exports, and permit civil flights. In short, smart sanctions focus on relaxing the movement of civilian population and products, while tiding control over military items sought by the Iraqi government. The objective is to prevent the Iraqi regime from rebuilding its WMD and regaining conventional military strength, while lessening the burden on the civilian population by easing the flow of commercial goods. This proposed policy would take the tool of sanctions as propaganda away from Saddam Hussein and improve Washington's image in the Arab world. In May 2001, a UK draft resolution, supported by the United States, containing the guidelines for a smart sanctions program was presented to the UN Security Council. After several weeks of intense debate, the United States and the United Kingdom withdrew their smart sanctions proposal rather than risk a threatened Russian veto. Still, Washington vowed to keep working on modifying and implementing these new smart sanctions.

Developments on the international scene since September 11, 2001, have paved the road for a compromise between Washington and Moscow on how to proceed with Baghdad. On November 29, 2001, the UN Security Council unanimously passed resolution 1382, which contains the seeds of a potential new international consensus on Iraq. This compromise resolution provides that the Security Council will adopt a new list of imported goods for Iraq that will require approval to make sure they are not used for military purposes. This is an integral part of the smart sanctions draft resolution proposed in May 2001 by the United States and the United Kingdom, which called for the free flow of civilian goods and services to Iraq. On the other hand, the United States agreed that the Security Council would clarify what steps were needed to lift the economic sanctions. Resolution 1382 provides a breathing space for all parties concerned. It allows the United States to focus on the war against the Taliban and al-Qaeda while building a regional and international consensus for the return of weapons inspectors to Iraq. It also gives Russia time to convince Baghdad to come to terms with a compromise formula that provides a more tangible link between the return of the inspectors and the suspension of sanctions.[35] Finally, Washington has made it clear that if Baghdad does not allow the inspectors back, all options are open.

A major prerequisite to implementing these new and modified sanctions is securing cooperation from Iraq's neighbors, particularly Jordan, Turkey, Iran, and Syria (where most of the smuggled oil is shipped). On the other hand, Iraq has threatened to punish these neighbors if they accept and cooperate with this new U.S. policy. Stopping Iraqi oil shipments would be particularly devastating to Jordan. When the Security Council was debating smart sanctions, the Jordanian government told Kofi Annan that it imports $750 million worth of oil a year from Iraq, its largest trading partner, and that 37 percent of Jordanian industries are dependent on trade with Baghdad.[36] If trade were cut off, Jordan's economy might face dire consequences. Turkey faces a similar but less severe scenario. These smuggled crude and petroleum products from Iraq are the economic lifeblood for southeastern Turkey. This region is still recovering from a long civil war between separatist Kurds and the Turkish government.

To sum up, even though Jordan and Turkey are close allies of the United States, promises of their cooperation in any modified sanctions regime should be taken cautiously. Given their historical and extensive relations with Baghdad, both Amman and Ankara are keen to avoid antagonizing the Iraqi government. Meanwhile, securing cooperation from Tehran and Damascus will be even more difficult given the cool and uneasy relations these two countries have with Washington. Still, despite these difficulties in persuading other countries to join the United States in modifying the sanctions regime imposed on Iraq since 1990, some analysts believe that removing the economic sanctions while containing Iraq militarily is the only workable policy short of waging a war.[37] Finally, it is important to point out that modifying and tightening international sanctions is one part of a broader U.S. strategy in dealing with Iraq. Two actions by which the Bush administration has sought to build on the accomplishments of previous presidents are containing Iraq's nonconventional military capabilities and increasing support to opposition groups that could precipitate a regime change in Baghdad.

Containing Iraq's Weapons of Mass Destruction

Washington perceives the proliferation of weapons of mass destruction (biological, chemical, nuclear, and ballistic) as a direct threat to American national interests and believes that Baghdad has been actively pursuing efforts to develop a stock of these weapons. An early exposure of the Iraqi program was made by the Israeli raid on the Osiraq reactor in 1981. The worldwide perception of Iran as a fanatic revolutionary power provided Iraq with an opportunity to develop its nonconventional weapon capabilities with few restrictions from the international community. At the end of the war against Iran, Saddam Hussein used chemical weapons against civilian and military targets in both Kurdistan and Iran. Thus, by 1990 it was known that Iraq had developed some of these nonconventional weapons. But outsiders do not know how advanced the Iraqi program is. The following analysis addresses three questions: What are Baghdad's motives to acquire and develop WMD? How has the United Nations tried to find and destroy Iraq's WMD capabilities? How does the United States assess Iraq's nonconventional capabilities?

Baghdad's efforts to acquire and develop WMD can be explained by the rivalry with three regional powers: Israel, Iran, and Saudi Arabia. Although geographically distant and lacking a common border with Israel, Iraq has almost always featured in Israeli thinking as an active, formidable confrontation state.[38] Indeed, the two countries have been technically at war since 1948, when Iraq participated in hostilities against Israel in the first Arab-Israeli war. Iraq was the only participant who refused to sign an armistice agreement with Israel in 1949 after the cessation of hostilities.[39] After a bloody military coup in 1958, the Iraqi monarchy was overthrown, and since then Baghdad has adopted a radical orientation in both domestic and foreign policies, particularly toward Israel.

Supported by huge oil revenues in the 1970s, Iraq emerged as a significant regional power and sought to acquire nuclear capability. In June 1981, Israeli planes launched a preemptive attack against the Iraqi nuclear reactor Osiraq before it went into operation.[40] This attack was also meant to maintain Israel's monopoly over nuclear weapons in the Middle East. (As early as 1970, it was universally assumed that Israel was a nuclear power.)[41] Israel, furthermore, is not a signatory to the Non-Proliferation Treaty. The deep animosity toward Israel in conjunction with Jerusalem's nuclear advantage has always been an important consideration in the Iraqi leaders' thinking and their quest for a strategic balance.

Iran probably represents a more serious challenge to Iraq than Israel. Since the establishment of Iraq as an independent state after the First World War, the leaders in Baghdad have not accepted the territorial boundaries with Tehran. What the Iraqis call "Shatt al-Arab" and the Iranians call "Arvand Rud" has always been a point of contention between the two countries and was a major reason for the 1980–88 war between them. This Iranian-Iraqi conflict has been dormant since the Gulf war. The two sides have not agreed on a final demarcation of their boundaries. Another sign of their rivalry is that each supports opposition groups against the other. In 1982, Iran helped to establish the Supreme Assembly for the Islamic Revolution in Iraq as an umbrella group for all the Iraqi Shi'ia parties. Meanwhile, Mujahedin-e Khalq, the main Iranian opposition organization, is headquartered in Baghdad and attacks Iranian targets from its military bases in Iraq. Finally, Iraq views the Iranian missile program with suspicion. In 1998, Tehran tested a new missile, Shahab-3, which can reach targets anywhere in Iraq.

In addition to the unresolved disputes with Israel and Iran, conflicts with Kuwait and Saudi Arabia provide another incentive for Iraq to seek WMD. For several decades, Iraq has claimed either parts of Kuwait (Bubiyan and Warbah islands) or the whole emirates as its own. In the early 1960s and 1970s, Baghdad threatened to use military force to attack Kuwait and was stopped by Arab mediation. In November 1994, Iraq formally accepted the UN-demarcated border with Kuwait, which had been spelled out in Security Council resolutions 687 (1991), 773 (1993), and 883 (1993). Iraq treats this official recognition, however, as unavoidable capitulation in the face of intense international pressure, and in all likelihood would subject it to revision if circumstances were to change. Furthermore, since the Gulf war, Kuwait has allowed American and British fighters to launch air strikes against Iraq from military bases there. The conflict with Kuwait is dormant and not yet settled to Iraqi satisfaction. The acquisition of WMD might help in a future confrontation.

Tension has characterized relations with another neighbor, Saudi Arabia. The old rivalry between the royal family in Riyadh, the al-Saud, and that in Baghdad, the Hashemite, continued until 1958 when a republican regime was declared in Iraq. This change in the Iraqi political system, however, did not pave the way for better relations with Saudi Arabia. Instead, the kingdom perceived the successor regimes in Iraq as radical opponents of its conservative pro-Western policy. The Iranian revolution of 1979 represented a threat to both Saudi Arabia and Iraq and helped them to create a tactical alliance to face the common challenge. This close cooperation faded shortly after the end of hostilities with Iran in 1988, and during the Gulf war Iraqi missiles landed in Saudi Arabia. In addition, in March 1988 it was revealed that the kingdom had covertly bought Chinese CSS-2 long-range ballistic missiles at a cost of $3 to $3.5 billion.[42] These missiles can reach targets in Iraq and are seen as a potential threat to the regime in Baghdad.

This proliferation of missiles and other kinds of nonconventional weapons as well as the perceived threats from Israel, Iran, and Saudi Arabia suggest that maintaining WMD is a matter of national security for the Iraqi government. Accordingly, Saddam Hussein has done everything he can to keep and upgrade his country's nonconventional weapons capabilities. A change in leadership is unlikely to weaken Baghdad's determination to acquire WMD. The rivalries with Jerusalem, Tehran, and Riyadh preceded the current regime and will continue after it. Put differently, regardless of who is in power in Baghdad, the Iraqi government will continue its quest for these weapon systems. A military analyst makes a similar argument: "Iraq is not proliferating simply because its current regime is radical and extreme; it is proliferating because it has strategic reasons to do so."[43]

In 1974, Iraq initiated an ambitious nuclear program in cooperation with France.[44] France agreed to supply Iraq with a large reactor, technical assistance, and training in return for substantial imports of oil from Iraq.[45] The destruction of the Osiraq reactor in 1981 did not weaken Baghdad's determination to become a nuclear power. Rather, Iraq sought to develop a uranium enrichment production capacity and heavy water.[46] Outside the nuclear arena, Iraq skillfully played throughout the 1980s on the West's fear of Iranian fundamentalism by presenting itself as a secular bulwark against the Khomeini regime. This universal perception facilitated a huge Iraqi military buildup, which included both conventional and nonconventional weapons. The nuclear revelations in the summer of 1991 came as a shock to the whole world. Iraq had clearly made more progress toward a nuclear bomb than anyone had previously suspected. According to David Kay, who led early UN nuclear weapons inspections in Iraq, "At the time of the Gulf war, Iraq was probably only eighteen to twenty-four months away from its first crude nuclear device and no more than three or four years away from more advanced deliverable weapons."[47] Besides these efforts to acquire nuclear capability, Iraq had developed an ambitious biological weapon (BW) program, about which very little was known until the defection of Hussein Kamel al-Majid, Saddam's son-in-law, in August 1995. The revelations shed new light on the magnitude of Baghdad's chemical, biological, and nuclear programs.

Against all odds, UNSCOM did an excellent job in carrying out its mission. In 1998, the UN secretary-general, Kofi Annan, confirmed that the inspectors had succeeded in destroying more weapons in Iraq than were destroyed during the entire Gulf war. However, Iraqi compliance with Resolution 687 and cooperation with UNSCOM decreased sharply over time as Baghdad came to realize that no degree of compliance would deter Washington from its chief objective: removing Saddam Hussein from power.[48] This growing mistrust between UNSCOM and the Iraqi government resulted in several showdowns, particularly when Baghdad claimed that there were too many American and British inspectors and when UNSCOM was denied access to presidential palaces.

In 1998, UNSCOM was put on the defensive when one of its inspectors, Scott Ritter, resigned, claiming that the U.S. government had penetrated the Commission. Ritter also acknowledged that while working for the UN, he had received intelligence about Iraq from Israeli government agencies.[49] A few months later, U.S. officials admitted that American spies had worked undercover on teams of UN arms inspectors, ferreting out secret Iraqi weapons programs.[50] Meanwhile, Kofi Annan confirmed that he had obtained convincing evidence that UN arms inspectors helped collect eavesdropping intelligence used in American efforts to undermine the Iraqi regime.[51] This was considered a violation of the UN rules, since the world organization should not participate in an operation to overthrow the leadership of one of its member states. At the same time, Israel had the means and motives to assist the arms inspectors. By July 1995, "Israel had become the most important single contributor among the dozens of UN member states that have supplied information to UNSCOM."[52]

These revelations substantially undermined the credibility of UNSCOM and added one more reason for the Iraqis to end any cooperation with the inspectors. In December 1998, all the inspectors were withdrawn from Iraq. Immediately after their evacuation, U.S. and U.K. fighter jets bombed Iraq in what is called Operation Desert Fox. Though far more massive than any previous strike against Iraq since the Gulf war, Desert Fox remained a limited operation. The campaign was ended after only four days to avoid adverse political and diplomatic consequences expected to arise if strikes continued during the Muslim holy month of Ramadan. Nevertheless, France, Russia, China, and Egypt all protested the Desert Fox bombing, and demonstrations were held in much of the Arab world.

Since then, no inspections have taken place. Iraq decided it would not cooperate with UNSCOM. The arms inspectors, however, still have not finished their job. The final report submitted by UNSCOM to the Security Council states that "in virtually every major weapons category, there are still outstanding 'disarmament issues' that merit further investigation."[53] After a comprehensive review of the UN approaches to Iraq, the Security Council adopted Resolution 1284 in December 1999. Sponsored by Britain and the Netherlands and supported by the United States, this resolution has two objectives: to revive weapons inspections and to improve the humanitarian conditions in Iraq. A new arms agency, known as the United Nations Monitoring, Verification, and Inspection Commission, was to replace UNSCOM. The resolution specified that Iraq should fulfill key disarmament tasks and demanded that Iraq grant the new inspectors "immediate, unconditional, and unrestricted access to any and all areas they wish to inspect."[54] At the same time, the resolution lifted the cap on the volume of oil Iraq can sell to pay for food, medicine, and other humanitarian supplies and relaxed some control on Iraq's imports. Baghdad rejected the resolution, arguing that it had already fulfilled all disarmament requirements and deserved an immediate lifting of sanctions.

In the aftermath of Operation Desert Fox, U.S. and British jets have engaged in limited strikes primarily against Iraqi air defense sites but also against communications nodes and other targets. This low-grade aerial bombardment against Iraq is designed partly to avoid public criticism and partly not to incite sympathy for Iraq among Arab and Muslim masses. As the years progressed, allies grew less supportive of military strikes on Iraq. Saudi Arabia began curtailing the types of missions that could be flown from its territory, and Turkey at times refused the United States permission to conduct strikes on Iraq. These strikes have contributed to the erosion of the consensus around inspections and sanctions. Shortly after the Bush administration took office in 2001, concerns regarding these aerial strikes were raised. Their maintenance is wearing down U.S. military equipment, straining U.S. resources in the region, and providing Saddam Hussein with valuable fodder for his propaganda war. Furthermore, the resumption of civilian flights within Iraq and international flights to Baghdad increase the possibility of airborne mishaps, which could carry significant collateral costs both in terms of human lives and U.S. standing in the Gulf.[55]

The U.S. government is seriously concerned about Iraq's current and potential capability to acquire and develop nonconventional military capability. The September 11 terrorist attacks and their aftermath have highlighted American vulnerability to biological warfare. Federal investigators have found no evidences linking Iraq to the anthrax letters. Still, the whole episode has added to Washington's determination to ensure that Iraq's weapons of mass destruction have been completely destroyed.

In the area of nuclear weapons, Washington believes that Baghdad would need five or more years and key foreign assistance to rebuild the necessary infrastructure to enrich enough material for a nuclear weapon. This period would be substantially shortened if Iraq acquired fissile material from a foreign source. With regard to the biological program, Washington believes that Baghdad has the ability to reconstitute its biological warfare capabilities within a few weeks or months. Furthermore, in the absence of UNSCOM inspections and monitoring, Iraq may have produced some biological warfare agents. Similarly, the U.S. government believes that since the Gulf war Saddam Hussein has rebuilt his chemical production infrastructure, including dual-use chemical warfare–associated production facilities, destroyed by the U.S. bombing during Operation Desert Fox. Finally, according to U.S. official documents, Iraq retains a limited number of ballistic missiles capable of striking its neighbors and has warheads capable of delivering chemical or biological agents.[56]

Four conclusions may be drawn from the foregoing discussion of Baghdad's nonconventional military capabilities. First, Iraq is not the only country to manufacture and develop WMD. Indeed, many countries, including some of Iraq's neighbors, already possess this capability. However, international efforts have focused on eliminating Iraq's WMD because Saddam's regime has used significant quantities of chemical weapons in previous armed conflicts against the Iranians and its own Kurdish citizens. Second, for several years Iraq has claimed that all its WMD have been destroyed and that sanctions should be lifted immediately. On the other hand, the United Nations inspectors who served in Iraq agree that the country's nuclear weapon program has been destroyed and that her substantial chemical arsenal has been largely neutralized. Still, chemical programs have some outstanding questions on quantities of poison produced, especially the nerve agent VX. Although it is not possible to exclude the existence of some missiles and launchers, Iraq's delivery capacity has also been sharply reduced. The greatest concern is in relation to biological weapons where the UN inspectors were never able to account for all the stocks of biological agents, which Iraq denied having until 1995, or for large amounts of growth material used to produce germ warfare agents.[57]

Third, regardless of the destruction of Iraq's WMD and the facilities to manufacture them, it is impossible to eliminate the know-how that had been developed by Iraqi scientists. Iraq has the knowledge to make these weapons, and this knowledge cannot be taken away. Finally, air strikes by themselves cannot destroy Iraq's WMD. In spite of all its shortcomings, UNSCOM did an excellent job in finding and destroying the Iraqi weapons. The absence of UN inspectors on the ground in Iraq since 1998 is a big setback to the international efforts to deny Baghdad nonconventional capabilities. Another option that has been considered by Washington for some time is to encourage and facilitate a regime change in Baghdad.

Supporting the Opposition

Since the Gulf war, several top American officials have expressed Washington's desire for a regime change in Baghdad. In December 1998, Secretary of State Madeleine Albright stated that the American policy of containment of Iraq had changed to one of "containment plus regime change."[58] Similarly, in 1998 Donald H. Rumsfeld and Paul D. Wolfowitz, who later became secretary of defense and deputy secretary of defense in the Bush administration, called on President Clinton to recognize a provisional government of Iraq. Wolfowitz went further by suggesting that the United States should be prepared to commit ground forces to protect a sanctuary in southern Iraq where the opposition could safely mobilize.

Since 1991, the United States has taken several steps to weaken and discredit and eventually overthrow the Iraqi regime. Shortly after the end of the Gulf war, President Bush called on the Iraqi people to "rise up and overthrow the dictator Saddam."[59] The largest sectarian and ethnic groups, the Shi'ites and the Kurds, perceived the president's statement as a promise of American support. In early March 1991, armed rebellion broke out among the largely Shi'ia population of southern Iraq, and almost simultaneously there was a similar rebellion by the Kurds in the northern part of the country. However, a fundamental shift in American policy had occurred. The Bush administration had concluded that preventing the disintegration of Iraq should take priority over the overthrow of Saddam Hussein. Noting Washington's hesitation, the Iraqi president used his military power to crush the two rebellions. In addition, Saddam Hussein has played up fears of Iraq's dismemberment to maintain domestic and international support for his regime.

In response to the Iraqi brutality in crushing the rebellions, the United States, Britain, and France established two no-fly zones: one in the north to protect the Kurds, and one in the south to protect the Shi'ites. Washington supported the two main Kurdish political parties—the Kurdistan Democratic Party (KDP) and the Patriotic Union of Kurdistan (PUK)—to establish an autonomous government. The hope was that a democratic experiment in the north could serve as a model for the rest of Iraq. The two Kurdish factions, however, have engaged in sporadic fighting since 1994. Worse, in late August 1996 the United States received a serious blow to its intelligence gathering when the KDP invited the Iraqi army to assist in the fighting against the PUK. The Iraqi army arrested and executed many of its opponents (e.g., American supporters).

In October 1998, President Clinton signed the Iraqi Liberation Act, which called on the administration to select one or more Iraqi opposition groups to receive as much as $97 million in Defense Department equipment and military training to overthrow Saddam Hussein. In the following months, the Clinton administration identified seven groups to receive American assistance: Iraqi National Accord (INA), Iraqi National Congress (INC), Islamic Movement of Iraqi Kurdistan, Kurdistan Democratic Party, Movement for a Constitutional Monarchy, Patriotic Union of Kurdistan, and Supreme Assembly for the Islamic Revolution in Iraq (SAIRI). Some of these groups, however, have been reluctant to be closely associated with the United States. In an effort to undermine the government of Saddam Hussein, the White House gave Iraqi opposition groups permission to resume their activities inside Iraq with American funding only a few weeks after George W. Bush took office. According to the plan, prepared in close consultation with the INC, opposition members will make clandestine forays into government-controlled areas to distribute relief supplies and propaganda.[60]

This policy, however, lacks consensus even within the U.S. government. General Anthony Zinni, who served as commander-in-chief of U.S. Central Command, responsible for operations in southern Iraq, warned that arming opposition groups, which he believes are ineffective and divided, could lead to a civil war similar to the experience in Afghanistan.[61] Others argue that such a move would be a replay of the American fiasco in the Bay of Pigs.[62] Similarly, Secretary of State Colin Powell has repeatedly expressed skepticism about the capabilities of the Iraqi opposition.

On the other hand, Arab countries, whose assistance would be crucial for the plan to work, have refused direct American intervention in toppling the regime in a fellow Arab country. They are not ready to host armed resistance against the government in Baghdad because they believe this would lead to a civil strife. Instead, their favorite option is a palace coup. Furthermore, Saudi Arabia is concerned that any opposition success might increase Shi'ia influence in Iraq and further complicate the delicate balance within the kingdom between the Sunni majority and the Shi'ia minority. Turkey, for its part, is suspicious of any plan that might augment the military capabilities and increase the political aspirations of the region's Kurds.

Confronted with these objections, the United States has been reluctant to provide arms and military training to the Iraqi opposition. Rather, under the Clinton administration, the State Department decided to supply the Iraqi opposition groups with nonlethal aid such as offices, fax machines, and other communication equipment, including Radio Free Iraq, as well as train them in civil administration. It is doubtful that these methods can topple Saddam Hussein. His hold on power is too strong and his opponents are too divided to create optimal conditions for a successful coup.

Saddam's relative success and his opponents' failure can be explained by three characteristics of the Iraqi political system. First, the opposition groups reflect the same division and fragmentation of the body politic in Baghdad. There are Kurdish, Shi'ia, Baathist, Arab nationalist, Communist, and liberal democratic opposition groups. They are united neither in their methods nor in their final objectives. Second, the foreign powers, which have interests in the ultimate future of Iraq, have not stood still. Many have been active on the political scene in Baghdad, sponsoring some groups to serve as their agents. In 1982, Iran helped to establish the SAIRI as an umbrella group for all the Iraqi Shi'ia parties under the leadership of Muhammad Baqir al-Hakim. In 1990, Syria participated in the creation of the Joint Action Committee, and Jordan played host to a number of the Iraqi opposition groups. The INA has its headquarters in Amman. Finally, the INC, a coalition of opposition groups, was formed in October 1992 and has since enjoyed financial support from Washington and Riyadh. These extensive and contradictory links with foreign powers have intensified the competition between factions and contributed to the ineffectiveness of the opposition as a whole. Third, the brutality of Saddam's regime has added to the structural weakness of the opposition. Most of these groups operate from outside the country with very limited roots inside Iraq. Furthermore, no national figure has emerged to unite the factions and to appeal to the general public.

Three conclusions can be drawn from the experience of the Iraqi opposition since the Gulf war. First, Saddam's regime is not likely to be overthrown by an outsider group or by a "hoped-for" popular uprising. If there is to be a change, it is more likely to come from the core of the regime, the Arab Sunni minority in Baghdad. Potential replacements are expected to be recruited from the army and the Baath Party. Second, any assessment of U.S. involvement with the Iraqi opposition should not be based solely on removing Saddam Hussein from power. Rather, the presence of such opposition has forced the Iraqi regime to use some of its assets for internal security. Third, rhetoric aside, the United States is not likely to vigorously support the Iraqi opposition. Caution at home and opposition by major powers and allies in the Middle East have led Washington to avoid an aggressive military intervention or sending massive arms supplies to the opponents of the Iraqi regime. A palace coup will continue to be the favorite scenario in Washington and elsewhere. This conventional wisdom, however, has been substantially shaken by the fast and decisive war against the Taliban (the so-called Afghani model).

Iraq since September 11

In the aftermath of the September 11 terrorist attacks, the question of what to do with Saddam Hussein has resurfaced. Although no credible evidence has been found linking Iraq to al-Qaeda or to the anthrax letters, many policymakers and journalists have speculated whether Iraq should be the next phase in the U.S. war on terrorism. This can be explained by two developments. First, as horrible as the September 11 terrorist attacks were, they could have been much worse if chemical weapons, biological agents, or nuclear devices had been used. Baghdad is believed to have some of these capabilities and strong resentment against the United States. Second, the rapidity with which the Taliban regime collapsed and a homegrown interim government was installed is seen by some in the political establishment in Washington as an example that can be applied in Baghdad.

Despite these developments, policymakers in Washington have been strongly divided on what action should be taken to neutralize the threats coming out of Baghdad. In early December 2001, Bush issued an ultimatum demanding that Saddam Hussein allow weapons inspectors back in Iraq or "find out" the consequences. Similarly, leading members of Congress[63] sent a letter urging Bush to make the Iraqi regime the next major target in the war on terrorism, declaring that "as we work to clean up Afghanistan and destroy al-Qaeda, it is imperative that we plan to eliminate the threat from Iraq."[64]

Besides leading members in the Congress, other proponents of a swift military action against Iraq include Rumsfeld, Wolfowitz, the president's counterterrorism chief, Wayne A. Downing, the vice president's chief of staff, I. Lewis Libby, former Central Intelligence Agency director James Woolsey, and former secretary of state Henry Kissinger. Their argument is based on several assumptions.

(1) The "Afghani model," an American air campaign using high-tech precise munitions and targeting by U.S. Special Forces on the ground in support of an indigenous armed opposition, can be replayed in Iraq.
(2) The war on the Taliban and al-Qaeda has generated a global momentum against regimes that sponsor terrorism.

(3) Saddam Hussein is so unpopular that a determined move should be enough to set his own people against him, and (as the experience in Afghanistan suggests) U.S. troops will be seen as liberators, not as an occupying force.

(4) There is no need for credible evidence linking Saddam Hussein to the September 11 terrorist attacks. Rather, his hatred for the United States and proven capacity to do something about it justifies military action.

As James Woolsey argues, "We need to destroy this regime that wants to destroy us and terrorize its neighbors. . . . Living with him is the most dangerous course to take."[65] Henry Kissinger reached a similar conclusion: "There is no possibility of a negotiation between Washington and Baghdad and no basis for trusting Iraq's promises to the international community."[66]

On the other hand, the opponents of immediate military action against Iraq as the second phase of the American war on terrorism include Secretary of State Colin Powell, his deputy, Richard L. Armitage, and the Bush administration's Middle East envoy, Anthony Zinni.[67] Their counterargument is also based on several assumptions.

(1) The overwhelming international support for the American war against the Taliban and al-Qaeda would fall apart if the United States took unilateral action to topple Saddam Hussein, especially since there is no evidence linking him to the September 11 terrorist attacks. Several U.S. allies in the war on terrorism have advised the administration against striking Baghdad. Such a military operation would also revive accusations of unilateralism against the United States.

(2) Despite some shortcomings of the sanctions regime, Saddam Hussein is basically being contained.

(3) Toppling Hussein's regime would open the door for several uncertain and dangerous possibilities including the disintegration of Iraq and disruption of oil supplies from the Persian Gulf region.

(4) There are fundamental differences between the war in Afghanistan and a potential military operation in Iraq. The Iraqi army is much stronger and better equipped than the Taliban, and the Iraqi opposition is weaker and not as well organized as the Northern Alliance in Afghanistan. As Powell argues, "They are two different countries with different regimes and different military capabilities. They are so significantly different that you can't take

the Afghani model and immediately apply it to Iraq." General Zinni argues along the same lines by describing an American military support to Iraqi opposition as a "prescription for a 'Bay of Goats' dreamed up by some silk-suited, Rolex-wearing guys in London."[68]

In the final analysis, the decision to take decisive military action to topple Saddam Hussein will probably depend on Bush's ability to convince the American public that another war against Iraq is a worthwhile cause and that Americans should be willing to pay the price in terms of economic costs and human lives. Similarly, the administration will have to convince U.S. allies in the Arab world, Europe, Russia, China, and others to continue supporting American military operations. Finally, another war against Iraq will depend on how Saddam Hussein responds to the international pressure to allow weapon inspectors back to Iraq and completely rid his country of weapons of mass destruction. The situation between the United States and Iraq is pregnant with several uncertain and dangerous possibilities.

Conclusions

At the end of the Gulf war, nobody imagined that more than a decade later Saddam Hussein would still be in office. Against all odds, he has not only survived the defeat but has consolidated his grasp on power, and his regime seems more stable than it was in the early 1990s. U.S. implicit or explicit policy to overthrow Saddam Hussein has not succeeded. Baghdad has paid a very high price for the prolonged and comprehensive sanctions regime imposed on the country since August 1990. This price, however, is being paid by the ordinary citizenry, not by the leadership. Similarly, the tremendous international efforts, led by the United States, to destroy Iraq's nonconventional military capabilities have achieved modest success. Very few believe that Iraq is completely free of WMD. Finally, a legitimate opposition group with roots inside Iraq and a workable plan to overthrow Hussein's regime is yet to be found. The underlying conviction in Washington is that the Iraqi regime has failed to comply fully and unconditionally with the United Nations resolutions and has given up none of its ambitions to be the dominant power in the Gulf or its determination to retain the basic capability to rebuild its arsenal of WMD.

This assessment, however, should not be taken as a complete failure of the American policy toward Iraq. At least two achievements deserve special attention. First, there is no denial that militarily Iraq would have been much stronger without the close scrutiny of its policy and resources enforced by the United States. Second, since the liberation of Kuwait, Baghdad has been contained and prevented from threatening and attacking its neighbors. Washington should get at least some of the credit for preventing another war in the Persian Gulf. Still, despite some success, it is clear that after more than a decade of economic and military containment the United States has not articulated a cohesive and long-term strategy toward Iraq. Such strategy should consider two factors. First, Iraqi reserves represent a major asset that can add substantial capacity to world oil markets. The country has eleven major giant oil fields awaiting proper development and a dozen lesser prospects as well as the nine blocks of the Western Desert still waiting to be explored.[69] In short, a fully explored and developed Iraq is critically needed to meet rising global demand for oil.

Second, the internal political dynamics in Iraq add to the uncertainty regarding the country's future and the prospects for its full reintegration in the regional and international systems. The modern history of Iraq does not provide one single example of a smooth or peaceful transfer of power from one ruler to another. Saddam Hussein may be gone tomorrow, but he just as easily might outlast another American president. Will one of his two sons (Odai and Qusai) succeed him? In addition to this uncertainty regarding the succession question, Iraq is notable for lack of political reform. Over the past decade, several countries that Washington used to call "rogue states" have shown some signs of reform and moderation. Iran has held free elections, and some of its leaders "talk" about starting a dialogue with the United States. President Qaddafi of Libya handed over the two suspected terrorists in the 1988 bombing of Pan-Am Flight 103 to face trial in Scotland. (One of them was indicted.) The younger Assad in Syria has taken some steps to reform his country's economic system and to reduce its role in Lebanon and is on speaking terms with the United States. Sudan supported the U.S. war on terrorism by sharing crucial intelligence information on al-Qaeda and Osama bin Laden. On the other hand, there is little, if any, sign of reform or moderation in the Iraqi policy in more than a decade.

To sum up, the Iraqi dossier is far from being closed. Washington has yet to find a way to reconcile the need to keep the Iraqi oil flowing into the global markets with its desire to deter Baghdad from committing any aggression against its neighbors or acquiring the means to pursue such a policy. Since the United States can neither engineer Saddam Hussein's fall nor accept him back into the international community, it really has one option left: trying to sharpen the existing policy of containment.

5

The United States and Iran

Prospects for Rapprochement

With a population approaching 70 million, the Islamic Republic of Iran is the most populous country in the Persian Gulf. This demographic advantage has made Tehran a major player in any security configuration. Long-term stability in the region is unlikely without Iran. Furthermore, the country possesses the world's second largest natural gas reserves (after Russia) and the fifth largest petroleum reserves (after Saudi Arabia, Iraq, the United Arab Emirates, and Kuwait). Finally, the Islamic Republic offers easy access to the exploitation of the Caspian Basin's energy resources and to neighboring Commonwealth of Independent States nations, both in the Caucasus and Central Asia. Thus Iran continues to be a major player on the global energy scene.

The Islamic revolution of 1979 represents a turning point in Iran's history. In a brief period, the clerical regime, founded by the late Ayatollah Khomeini, succeeded in establishing itself and was able to survive, overcome, and adjust to several daunting internal and external challenges. These include the eight-year war with Iraq, the assassination of prominent revolutionary leaders, the death of Ayatollah Khomeini, and the American economic sanctions. The Islamic regime has been able to weather all these storms. President Muhammad Khatami, who was elected in 1997 and reelected in 2001, has sought to introduce and consolidate economic and political reforms. Given the central role that oil revenues play in the Iranian economy, the energy sector has been the focus of this reform. The process of opening up Iran's upstream oil and gas sectors, however, has been complicated by the hostile relations with the United States.

This chapter is divided into three parts. The first part reviews the history and evolution of Iran's energy industry. Particular attention is given

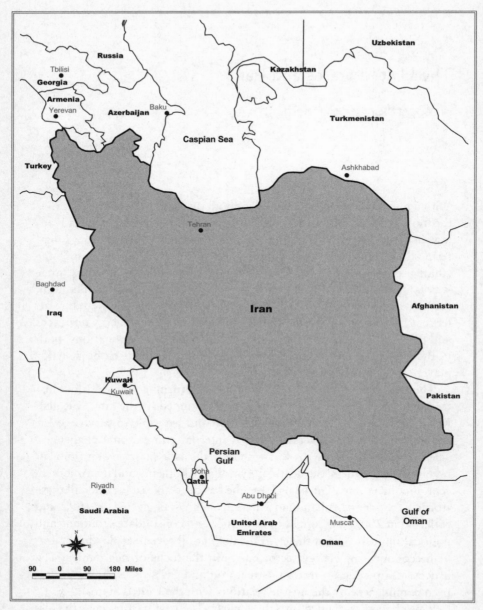

Map 5. Iran

to the recent efforts to invite international oil companies back to invest in the exploration and development operations of oil and gas fields. This leads the analysis to discuss the attempt by the United States to deny the Islamic republic access to foreign investment (e.g., American economic sanctions). The second part examines the main strategic issues that have complicated the relations between Washington and Tehran since 1979. These include Iran's involvement in international terrorism, attempts to acquire and develop weapons of mass destruction, and opposition to the Arab-Israeli peace process. The September 11, 2001, terrorist attacks and the subsequent war have provided a window of opportunity for the two countries to explore chances for cooperation. The potential strategic impact of these events on the American-Iranian relations will be analyzed in the third part.

Despite these impediments, Washington and Tehran share fundamental national interests. A stable and prosperous Iran can play a positive role in stabilizing global energy markets and promoting regional stability in the Persian Gulf and the Middle East. The outcome of the rivalry between moderate and extremist forces within the religious/political establishment in Tehran will have a significant impact on the speed and magnitude of a rapprochement with Washington. The sooner the two countries can articulate a policy of détente, the better for them, the international economy, and global peace.

The Oil and Gas Industry in Iran

The discovery and development of oil in Iran reflect three distinctive characteristics: Iran was the first country in the Persian Gulf and the Middle East where oil was found; the relations between the Iranian authorities and international oil companies in charge of petroleum exploration and development operations within the country were, generally speaking, characterized by mistrust and tension; and U.S. involvement in the Iranian hydrocarbon industry started in the mid-1950s, more than four decades after oil was discovered.

First, in 1901 the shah of Iran granted a concession to William Knox D'Arcy, a British adventurer, to find, exploit, and export petroleum anywhere in Iran, except for the five northern provinces (Azerbaijan, Gilan, Mazanderan, Astrabad, and Khorassan), which were excluded as a result of Russian influence. Oil was first struck in 1908 in Masjid-i-Suleiman, on the site of an ancient fire temple. The Anglo-Persian Oil Company was formed in 1909; renamed the Anglo-Iranian Oil Company in 1935, the company is now known as British Petroleum. Britain was particularly interested in oil discovery in Iran for at least three reasons: (1) there were no indigenous oil deposits in the United Kingdom nor in any part of the British Empire, so far as was then known; (2) in 1913, shortly before World War I began, the British Admiralty, then headed by Winston Churchill, decided to shift from coal to oil as fuel for the Royal Navy;[1] and (3) British control of Iranian oil would strengthen a British presence and influence in the Middle East and deter the threat of German or Russian expansion in that area.[2] Taking these reasons into consideration, the British government bought a controlling interest in the Anglo-Persian Oil Company. Accordingly, Iranian oil facilities were rapidly expanded during World War I and by the early 1950s were still the best developed in the Persian Gulf region.

Second, in spite of the continuing expansion of the Iranian oil industry, the tension and suspicion between the Anglo-Iranian Oil Company and the authorities in Tehran were building to a showdown in the early 1950s. The Iranian grievances focused on three areas: (1) the monopoly position enjoyed by the company; (2) the close relationship between the company and the British government; and (3) dissatisfaction with the financial terms of the concession between the company and the Iranian government. Developments between the early 1930s and the 1950s had underscored these Iranian complaints and contributed to the rise and sharpening of nationalism in Tehran. In 1932, the Iranian government decided to cancel the concession it had awarded to the Anglo-Persian Oil Company three decades earlier. After lengthy negotiations, the two sides signed a new concession in 1933 with more favorable provisions to Iran. Following this new agreement, the Iranians were satisfied with the steadily increasing income they were receiving and for a number of years had good relations with the company. However, the Second World War introduced new dynamics. In 1941, Britain and the Soviet Union occupied Iran and forced Shah Reza, who was sympathetic to Germany, to abdicate in favor of his son. Several months later, Tehran, London, and Moscow signed a tripartite treaty of alliance. In 1948 and 1949, negotiations between the Iranian government and the Anglo-Iranian Oil Company produced an agreement supplementary to the 1933 concession, offering improved financial terms to Iran, though allowing little scope for the assertion of Iranian sovereignty over its oil resources.[3]

The Iranian Majlis rejected the agreement and passed the Nationalization Bill in April 1951, providing for the creation of a National Iranian Oil Company (NIOC), to which the assets of the Anglo-Iranian Oil Company were to pass and which would be the government agency responsible for running all aspects of the Iranian oil industry. This movement was led by Dr. Muhammad Mossadeq, who became prime minister. In response to these changes, Iranian oil production dropped steeply and alternative oil resources were rapidly developed in the Persian Gulf, particularly in Kuwait, Saudi Arabia, and Iraq. The next two years witnessed an economic confrontation between London and Tehran, with Washington fearing that popular discontent and economic straits might lead Iran toward communism. These economic and political uncertainties came to an end by a coup d'etat, supported by the U.S. government, that led to the arrest of Mossadeq and the installation of a friendlier government in Tehran. Later, an agreement was signed between the new Iranian government and the Iranian Oil Participants, which was generally known as the Consortium. The Consortium was initially made up of several international oil companies: British Petroleum, 40 percent; Royal Dutch/Shell, 14 percent; Gulf, Mobil, Standard of New Jersey, Standard of California, and Texaco, each 7 percent; Compagnie Francaise des Petroles, 6 percent; and Iricon Agency, comprising eight U.S. independent oil companies, 5 percent.[4] The terms of the agreement were in line with the offers that had been made to, and refused by, Mossadeq at various times after the act of nationalization. Exploitation and marketing rights were assigned to the Consortium, but it was recognized that the NIOC remained the sole owner of all fixed assets. Thus, in principle, the 1954 agreement recognized Iranian nationalization of the oil industry and provided what amounted to a 50/50 share of profits. In 1966 the Consortium relinquished 25 percent of its agreement area, and in 1973 the NIOC assumed control over oil production and refining in the whole agreement area.

Third, in addition to the gradual American involvement in the Iranian oil industry, which started with U.S. oil companies' owning 40 percent shares of the Consortium agreement signed in 1954, Washington has had strategic interests in Tehran as a bulwark against Soviet expansion in the Middle East and west Asia. American influence in Iran was negligible until the early 1940s. During World War II, American troops and technicians went to Iran at the request of the British to assist in forwarding supplies to the Soviet Union to assist it in the war against Germany. Shortly after the war, the British role in helping Iran to resist Soviet pressure was taken over by the United States. The Truman Doctrine, which pledged

U.S. assistance to small nations against communist aggression and infiltration, was enacted in 1947. In line with this policy, specific economic and military aid was pledged and given to Turkey and Greece, and subsequently the policy was extended to Iran. Thus, between 1945 and 1961, the United States gave Iran $1.1 billion in foreign aid, most of it for economic assistance.[5] Meanwhile, military cooperation between the two nations had substantially increased until the 1979 revolution, particularly in the areas of arms sales and training. An important step in this alliance between Tehran and Washington was taken in October 1955 when Iran joined the Baghdad Pact, of which Britain, Turkey, Iraq, and Pakistan were already members and with which the United States was closely associated.

To sum up, the history and evolution of the Iranian oil industry are quite different from those of other states in the Persian Gulf. Oil was found in Iran earlier than the rest of the region, and the United States became involved in the Iranian oil industry in a later stage than the Saudi oil industry, which was developed from the beginning entirely by American oil companies. Finally, the process of expanding Iranian control over and ownership of its oil resources took place earlier and was more confrontational than the experiment in other Persian Gulf producers. In spite of Iran's failure in the early 1950s to completely control its hydrocarbon wealth, by 1973 Tehran took charge. Still, the NIOC allocated areas for exploration and development under joint arrangements with international oil companies and granted them service contracts.

Reopening Iran's Oil and Gas Sectors to Foreign Investment

The 1979 revolution was a turning point in Iran's overall development particularly with regard to the hydrocarbon industry. Almost all cooperation between the Iranian government and IOCs was frozen. This can be explained by two interrelated reasons. The new leaders in Tehran were extremely suspicious of foreign investors in general and especially IOCs. The dominant view was that a main goal of the revolution was to liberate the Iranian people and nation from foreign penetration, culturally, economically, and politically. On the other side, the political turmoil, which accompanied the revolution, followed by the war with Iraq (starting in 1980) made Iran a high-risk place in which to invest. In short, both sides—the Iranian authorities and IOCs—rejected any cooperation with the other.

By the late 1980s and early 1990s, however, the economic and political environment in Tehran had drastically changed as well as the country's stand on the regional and international scenes. The war with Iraq ended, Ayatollah Khomeini died, and several million people were added to the country's population, aggravating economic problems, particularly unemployment. By the early 1990s Iran was poorer than what it was before the revolution. This decline in the standard of living for the majority of Iranians has incited the leadership to adopt a less dogmatic and more pragmatic attitude toward foreign investment. Meanwhile, the responsible stand Tehran took during the Gulf crisis (1990–91) has improved its international image and facilitated its reintegration in the global economic system. Accordingly, since the mid-1990s the Iranian government has pursued orchestrated efforts to attract foreign investment. This is most apparent in the hydrocarbon sector.

Besides the growing realization that foreign investment is a necessity to address the fundamental imbalances in the Iranian economy, there are specific incentives to invite IOCs back to the oil and gas sector. First, as has been discussed earlier, Iranian oil fields are the oldest in the Persian Gulf region. These aged fields suffer from a depletion rate of 250,000 to 300,000 barrels per day (b/d), according to the Iranian Oil Ministry.[6] At the same time, the Iranian government has set an output target of 5.6 million b/d in 2010 and above 7.3 million b/d in 2020.[7] This would suggest a sustainable capacity of about 8 million b/d, twice the estimated capacity of 4 million b/d in 2002. Expanding output capacity is important to Iran in order not to lose its share in OPEC's production and to keep its credibility as a major oil producer and exporter. Second, production from offshore oil fields requires more financial resources and more sophisticated technology than onshore production. IOCs are in a better position than the Iranian government to initiate and accelerate the exploration and development of these offshore oil fields. Third, Iran will not be able to maintain its current level of production or expand it without substantial investments. Some estimates put Iran's required investment to reach the 7 million b/d level at approximately $30 billion.[8] Tehran simply does not have this money. Foreign investment is a necessity.

To sum up, Iran has neither the required financial resources nor the technological expertise to upgrade and modernize its aging oil fields. Accordingly, Iranian oil production reached its peak in 1974, and for almost three decades it has been way below this level, as shown in table 5.1.

The figures show the severe impact of domestic political instability and the war with Iraq on Iran's oil production. Since the early 1990s, there has

Table 5.1. Iran's oil production, 1970–2000 (thousand barrels per day)

Year	Production	Year	Production	Year	Production
1970	3,829	1981	1,380	1992	3,429
1971	4,540	1982	2,214	1993	3,540
1972	5,023	1983	2,440	1994	3,618
1973	5,861	1984	2,174	1995	3,643
1974	6,022	1985	2,250	1996	3,686
1975	5,350	1986	2,035	1997	3,664
1976	5,883	1987	2,298	1998	3,634
1977	5,663	1988	2,240	1999	3,557
1978	5,242	1989	2,810	2000	3,719
1979	3,168	1990	3,088	—	—
1980	1,662	1991	3,312	—	—

Source: Energy Information Administration, *International Petroleum Monthly* (Washington, D.C: Government Printing Office, June 2001), 38.

been a steady rise in output. In order to fully utilize its hydrocarbon resources, the Iranian government has embarked on serious efforts to attract foreign investment back to its upstream oil and gas industry. Four characteristics of the Iranian opening strategy can be identified:

(1) Originally, the focus was on developing natural gas fields, but more recently IOCs have been invited to participate in developing oil fields.
(2) Iran has articulated a formula called "buy-back" in its dealings with foreign companies.
(3) Given political sensitivities, IOCs were invited to develop offshore fields, but later they became involved in the development of onshore fields as well.
(4) U.S. economic sanctions against investment in the Iranian energy sector have complicated the process of rapid development of oil and gas resources.

Natural Gas

Iran holds 15.3 percent of the world's proven natural gas reserves (second only to Russia). In spite of these huge reserves, the country's share of world production is only 2.5 percent.[9] This large gap between actual production and proven resources and capacity reflects the current status of the gas industry in Iran—largely underutilized. Traditionally, Iran has been an oil producer and exporter for most of the past century. For a long time gas received low priority because oil offered better and higher foreign exchange returns. Still, gas exports began on a large scale in 1970 to the former Soviet Union.[10] These exports continued for several years until they were stopped due to price and payment difficulties. Since the early 1990s, official attention has focused on developing gas resources. This was partly in response to the skyrocketing of domestic oil consumption, which cut into crude exports and threatened to deprive the state of a substantial proportion of annual revenues. The Iranian government, under President Ali Akbar Hashemi Rafsanjani, responded by launching a gasification program in 1992. Under this program, hundreds of cities and towns were given access to gas. Consequently, domestic consumption of natural gas more than doubled: from 22.7 billion cubic meters (bcm) in 1990 to 62.9 bcm in 2000.[11] Thus, one analyst rightly concluded that the surge of natural gas in the domestic energy market was the salient feature of the 1990s in Iran's energy picture.[12] Indeed, by 2001 natural gas accounted for around 44 percent of Iran's total energy consumption.[13] Increasing gas production can serve at least four objectives: (1) to satisfy growing domestic demand; (2) to use as a source of foreign revenues by exporting gas to countries such as India, Japan, and Turkey; (3) to inject into depleted oil fields to maintain reservoir pressure and allow higher recovery; and (4) to free more oil for export. Indeed, some analysts estimate that a partial switch to gas use will leave an additional $2 billion of oil products for export a year.[14]

In order to achieve these goals, much of the new gas field development of recent years has centered on South Pars, one of the world's biggest offshore gas fields. South Pars, first identified in 1988, is an extension of Qatar's North Field. It accounts for about a third of total proven gas reserves and, not surprisingly, is the country's largest energy project.[15] The field is being developed in twenty-five phases. It is estimated that it costs between $850 million and $1 billion to develop each phase.[16] In addition to South Pars, there have been significant discoveries. In April 2000, the Oil Ministry announced the discovery of Iran's biggest onshore sweet gas field, Tabnak in the southern part of the country. It contains reserves of 15.7 trillion cubic feet of gas.[17] The amount is not particularly significant for Iran, but the field is very attractive because the gas is sweet and will require minimum investment to develop. In August 2000, another important sweet gas field, Homa, was announced. It has recoverable reserves of 133,100 million cubic feet (mcf).[18]

These massive natural gas reserves have incited the authority in Tehran to develop a national distribution network. The backbone of this grid is the Iranian gas trunkline network (IGAT). The idea of developing a national gas network has been around since the mid-1960s, when IGAT-1 was built, followed shortly by IGAT-2.[19] In order to meet a growing domestic demand, construction of IGAT-3 started in the early 2000s, and plans for IGAT-4 are under consideration.[20] All these trunkline networks carry gas from southern Iran, where major gas fields are, to the north, where most population centers are located.

Since the late 1990s, Iran has articulated a sound strategy to expand its natural gas output to meet growing demand both at home and in neighboring countries. Iran will probably emerge as a major gas producer and exporter. Indeed, the groundwork is being laid for major export projects, particularly to Turkey and India. Securing markets for gas, however, is different from securing markets for oil. Unlike oil, the process of transporting gas from the fields to the markets is expensive and difficult. Mainly, there are two ways to transport gas, either by pipelines to neighboring markets or as liquefied natural gas (LNG) to distant ones. Approximately a quarter of internationally traded gas, is shipped as LNG.[21] Both methods are characterized by large capital costs, relatively long lead times, and long-term commitments between consumers and suppliers, with "take-or-pay" twenty-year or longer contracts as the norm. Still, worldwide trade in natural gas is increasing as advances in technology lower the costs of long-distance pipelines and of LNG processing and transportation. Nonetheless, compared with oil, natural gas has less of an international market and is more sensitive to regional changes. Iran plans to use pipelines to carry gas to Turkey and LNG to supply gas to India.

Turkey is seen as an attractive market primarily for two reasons. First, with one of the largest populations in the Middle East (over 65 million) and a growing economy with very little indigenous gas resources, Ankara's dependence on and need for imported gas is rapidly rising. Second, Turkey is viewed by natural gas suppliers not only as a potentially booming gas market in its own right but also as the gateway to other consumers in southern and central Europe. In other words, Turkey's strategic location makes it a natural "energy-bridge" between energy producers in the Middle East and Caspian Basin regions on one hand and consumer markets in Europe on the other hand.[22]

With a share of total world gas consumption of only 0.6 percent in 2000, Turkey is not a major consumer. Turkish natural gas demand, however, is projected to increase rapidly in coming years, with the prime consumers expected to be power plants and industrial users. Natural gas is

Turkey's preferred fuel for new power plant capacity for several reasons: environmental (gas is cleaner than coal or oil), geographic (Turkey is close to huge amounts of gas in the Middle East and Central Asia), and economic (Turkey could offset part of its energy import bill through transit fees it could charge for oil and gas shipments across its territory).[23] This projected rising demand will be met by additional imports. Currently, the bulk of Turkish gas demand is met by imports from Russia, Algeria, and Nigeria. For the past several years, the Turkish government has negotiated with several sources for further imports to close the envisaged gap between projected demand and contracted supply. These include huge and expensive schemes to import gas from Turkmenistan (Trans-Caspian Pipeline), Russia (Blue Stream), and Azerbaijan (Shah Deniz).

Finally, in 1996 Iran and Turkey signed a $20 billion agreement that calls for Iran to supply Turkey with natural gas for twenty-two years. Exports of Iranian gas to Turkey were slated to start in 1999 at an initial rate of 300 million cubic feet a day (mcf/d) and rise to a level of 1,000 mcf/d in 2005.[24] Gas pumping, however, was delayed several times due to the failure to complete the construction of the necessary pipelines and a metering station. Furthermore, Turkey's economic slowdown in the early 2000s may have contributed to the postponement of the project. Finally, American opposition to the deal may also have played a role in Ankara's reluctance to begin importing gas from Iran. After a long delay, delivery of natural gas from Iran to Turkey started in December 2001.

Since the early 1990s, three developments have provided incentives for closer Indian/Iranian cooperation. First, after several decades of implementing a state-led economic policy, New Delhi started liberalizing its economic system and opened it up to foreign trade and investment. This shift in economic policy resulted in a significant growth in the country's gross domestic product. India will need increasing volumes of energy to sustain its rapid economic growth. Domestic gas supply will not keep pace with domestic demand. Thus, India is projected to grow more dependent than ever on imported gas. Most of this gas is likely to come from the Persian Gulf. Indeed, India has already either signed or negotiated gas deals with Qatar, Oman, the United Arab Emirates, and Iran. Second, New Delhi's relations with Arab countries have, to some extent, shifted focus. In the heady days of nonalignment, Egypt was the biggest partner India had in the Arab world. Iraq was another important ally, due to similarity in foreign policy orientation and mutual animosity toward Pakistan.

But the post–Cold War decade has seen Indian-Egyptian ties diminish for lack of economic incentives to sustain them. And since 1990, Iraq has been isolated by international sanctions. That has prompted India to increasingly turn its attention to the Gulf states and Iran. Third, the rise of the Taliban in Afghanistan provided strong political cement for the growing Indian-Iranian partnership. Both New Delhi and Tehran opposed the Taliban's policy and the support the movement received from Islamabad. Together these three developments—India's expanding demand for energy, the shift in its foreign policy, and the rise of the Taliban (1996–2001)—have contributed to the consolidation of relations between New Delhi and Tehran.

The problem Iran and India face is how to physically move large amounts of energy. The two countries have signed several memoranda of understanding and agreements regarding their energy cooperation. Several options have been under consideration including an overland or a deep-sea pipeline. Both options are challenging. An overland pipeline would have to pass through Pakistan. That raises security issues for India, which has fought three wars with Pakistan since the two countries won independence in 1947. While a deep-sea pipeline would be politically easier, it would also be much more costly and technically difficult than an overland line.

Iran, India, and Pakistan could yet find a way to ease the Indian-Pakistani rivalry sufficiently to participate in an overland pipeline. Pakistani officials estimate a pipeline could earn their country hundreds of millions of dollars in transit fees annually, as well as help Pakistan meet its own projected future energy shortages. Furthermore, the proponents of an overland pipeline say the presence of multinational investors in any construction would help to guarantee its security by making all players view it in international, rather than strictly national, terms. The destruction of the Taliban regime in late 2001 has intensified the Iranian efforts to start this huge scheme for two reasons. One, the Taliban constituted a major point of tension between Tehran and Islamabad, and their removal is likely to smooth relations between the two countries. Second, the establishment of a new government in Kabul is likely to contribute to political stability and make Afghanistan a potential route to export gas from Turkmenistan (not Iran) to Pakistan and India. Thus Iran, India, and Pakistan have formed joint technical committees to negotiate the details for a pipeline. Indeed, a pipeline stretching from Asaluyeh on Iran's Gulf coast to Pipavav in Gujarat state in northwestern India has been proposed.[25]

The United Kingdom's BG International has been involved in this scheme. Meanwhile, proposals to transport natural gas on LNG vessels from Iran to India are under consideration.

In addition to Turkey and India, Iran continues to seek export markets for its natural gas. Possibilities include Armenia,[26] Ukraine, Europe (via either Ukraine or Turkey), Pakistan, Taiwan, South Korea, and China. Finally, in July 2000 a memorandum of understanding was signed with Kuwait to study possible supplies of Iranian gas.[27] In order to expand its gas output to meet domestic demand and target these markets, Iran has signed several deals with international companies to participate in the exploration and development of its hydrocarbon resources.

Deals to Develop Offshore and Onshore Fields

Since 1995, Iran has embarked on serious efforts to attract IOCs to invest in the exploration and development of its hydrocarbon resources. An important step toward this was taken in 1995, when a list of eleven projects was offered to IOCs. The plan gained in scope and momentum in mid-1998, when twenty-four more oil and gas development projects, along with seventeen onshore and offshore exploration blocks, were put on offer.[28] This new Iranian attitude toward foreign investment in the country's energy sector has presented IOCs with a magnificent opportunity that they do not intend to miss. Several agreements between Iran and European, Canadian, Japanese, and other Asian companies have been signed recently.

In early 1999, Elf Aquitaine of France and Eni of Italy agreed to spend $540 million to increase production at the Doroud offshore field. Norsk Hydro has been awarded an exploration contract near the Iraqi border. Another Norwegian firm, Norex, is in charge of a project called Persian Gulf Carpet 2000, which requires seismic surveys in the Gulf. Similarly, the Royal Dutch/Shell Group has been involved in many projects in Iran, including the exploration in the southern Caspian Sea (in association with the London-based Lasmo), the development of Soroush and Nowruz oil fields, and a proposal for a gas-to-liquids plant. The UK's BG International has signed a joint venture agreement with an Iranian company aimed at developing Iran's first liquefied natural gas export project. Another British company, Enterprise, will participate in developing the South Pars. The biggest single award since the 1979 revolution ($3.8 billion) was given to the Italian firm ENI to develop the South Pars. Also, in June 2000 ENI signed a $900 million contract to develop the Dharkovin oil field.

Finally, an international consortium of nine oil and gas firms, mainly European, agreed to carry out a study for a twenty-five-year gas-utilization plan centered on supplies from South Pars.[29]

Unlike Europe, Japan has been hesitant to forge any close cooperation with Iran for fear of U.S. retaliation. In other words, both the Japanese government and Japanese companies are treading carefully for fear of upsetting their chief ally and biggest trade partner, the United States. Thus, under U.S. pressure, Tokyo in the mid-1990s suspended loans promised to Tehran. Since 1999, however, there have been signs of a growing Iranian-Japanese partnership, driven by at least four economic/strategic factors. First, Japan imports almost all of the oil it needs, and approximately 85 percent comes from the Gulf, particularly the United Arab Emirates, Saudi Arabia, and Iran. Tokyo is determined to maintain good relations with its oil suppliers. Second, in February 2000, a private Japanese firm, Arabian Oil Company, lost its forty-year rights to produce oil in the Saudi portion of the Neutral Zone oil field shared by Saudi Arabia and Kuwait. Thus, with Japanese firms effectively frozen out of oil development and production activities in Saudi Arabia and Kuwait, it seems that the Japanese government has decided to settle on Iran as its main prospective partner as far as upstream oil and gas development in the Middle East is concerned. Third, Japan is also eager to improve ties with Iran because of concern that China will increasingly compete for Iran's oil and win business contracts there. Fourth, Japanese companies fear they may be left behind by European, Chinese, and even U.S. companies poised to make contracts with Iran when the sanctions come to an end.

President Khatami's visit to Japan in late 2000 can be seen as a culmination of the emerging cooperation between Tehran and Tokyo. One of the results of this visit was a protocol under which Tehran agreed to give Japanese firms negotiating rights for the appraisal and development of a specific area of the newly discovered giant oil field, Azadegan, in southwestern Iran.[30] This oil field was discovered in 1999 and is considered the most important discovery for more than three decades. In addition, the NIOC pledged to support the participation of Japanese companies as minority partners in the development of other oil and gas fields.

To sum up, Iran has succeeded in attracting IOCs to invest in its energy sector. Since 1995, when Iran officially invited foreign investors to participate in the development of its oil and gas fields, the two sides signed deals worth more than $10 billion. All indications suggest that Tehran will seriously continue its efforts to negotiate and sign more agreements with

IOCs. Iran, however, needs to address two important obstacles: modifying the buy-back model and overcoming the U.S. sanctions.

"Buy-Back" Model

Following the 1979 revolution, a new constitution was established. Reflecting one of the issues at stake in the revolution itself, the new constitution stipulates that foreign ownership of the country's natural resources was illegal. Specifically, it prohibits the granting of petroleum rights on a concessionary basis or direct equity stake. The petroleum industry was nationalized, the government expropriated the assets of foreign petroleum companies operating in the country, and concession agreements were broken. Eight years later, when the attitude toward foreign investment started changing, the Petroleum Law of 1987 was promulgated. It permits the establishment of contracts between the Ministry of Petroleum and state companies on one side and local and foreign natural citizens and legal entities on the other side.[31] Since reopening to the IOCs, Iran has employed the buy-back framework for upstream contracts. The adoption of this principle represents a radical break with the past and was a courageous move that carried huge political risks for those who espoused it.

The buy-back model is better explained in comparison with other formulas that have been utilized between oil producers and foreign companies. These include joint venture, production-sharing agreement (PSA), and technical services contract (TSC). Joint ventures are arrangements under which a state enterprise and a foreign firm invest stated amounts of capital that can take various forms, including funds, intellectual property, or physical assets and rights to land. The partners share the risk and the reward of the venture in proportion to the capital contributed.[32] Under PSAs, the foreign contractor is reimbursed for exploration, development, and operating costs by way of a certain share of production.[33] In other words, the foreign partner holds no equity stake but does maintain a firm legal entitlement to a certain percentage of oil output volume. Finally, TSCs involve a simple cash payment for services rendered. These types of contracts entail little risk, if any, for the foreign firms. Subsequently, the rewards do not rise should discovery be substantial or oil prices increase.[34] In addition, service contracts tend to be awarded for short periods.

The buy-back formula does not violate the Iranian constitution. It is a service contract under which one or more parties are contracted by the Ministry of Petroleum to carry out necessary exploration and development work on a field that, once completed, reverts wholly to the ministry.

Thus, the foreign company is neither a partner nor a concessionaire, but acting in the role of a hired contractor servicing the national company. The buy-back model demands that the foreign partner provides all investment capital for exploration and, in return, is paid a predetermined rate of return on capital invested. This is paid in kind after the production of the first commercial oil. Buy-back contracts generally are designed by the Iranian negotiators to last five to seven years and are thus fairly short term in the context of the traditional upstream contract.[35]

Although there have been several agreements signed between Iran and IOCs based on buy-back formula, the model has been criticized particularly for three reasons. First, it is almost risk-free for the IOCs, since they are guaranteed a return on their investment. Second, the brief duration of the contracts does not provide incentives for the IOCs to introduce and fully utilize their advanced technology and management skills. Third, the fixed rate of return leaves IOCs with little incentive to maximize the profit from oil and gas fields. In short, the criticism focuses on the disconnection between performance and profit and the absence of long-term relationships between the Iranian oil authority and IOCs.

With these shortcomings taken into consideration, there have been intense debates in Tehran and negotiations with IOCs regarding modifying the buy-back model. In 2000 the Council of Guardians decided that the buy-back model supported a form of interest forbidden by Islam. This decision, however, was overruled by the Expediency Council, Iran's highest legislative court of appeal, in February 2001.[36] New contracts signed with IOCs link performance to payments and provide for long-term relations with foreign companies. The goal is to optimize oil and gas field utilization. A consensus is growing in Iran that, without modifying the buy-back formula, the country could fall short of its expectations in terms of attracting the necessary foreign investment. This is particularly important to offset how some foreign investors see Iran as a political risk due to the American sanctions.

U.S. Economic Sanctions

The United States has maintained sanctions against Iran since 1979, following seizure of the American embassy in Tehran. Economic sanctions against Iran became more exclusive in the mid-1990s with the signing of several executive orders and the enacting of the Iran-Libya Sanctions Act (ILSA). In 1995, President Clinton issued two executive orders that established a total ban on trade with Iran. The first order prevents U.S. companies, but not their foreign subsidiaries, from supervising, managing, and financing projects relating to the development of Iran's oil and gas resources. A second executive order stated that U.S. persons may not trade in Iranian oil, finance, broker, approve or facilitate such trading, or finance or supply goods or technology that would benefit the Iranian petroleum sector.[37] Even executory contracts whose legal effect is contingent upon the lifting of sanctions are not permitted.[38] As a result of these executive orders, the American oil company Conoco was obligated to abrogate a $550 million contract to develop Iran's offshore Sirri A and E gas fields.

To tighten the embargo further, Congress unanimously passed the ILSA, and President Clinton signed it into law in August 1996. ILSA mandates the president to impose two sanctions selected from a menu of six on any U.S. or foreign person who invests $20 million or more in an Iranian project (lowered in August 1997 from $40 million) if the investment directly and significantly contributes to the enhancement of Iran's or Libya's ability to explore for, extract, refine, or transport by pipeline its oil and natural gas. The menu includes denial of U.S. export licenses, restrictions on imports to the United States, a ban on U.S. government procurement of goods and services, a prohibition on certain loans or credits by U.S. financial institutions, restrictions upon sanctioned financial institutions, and a denial of Export-Import bank assistance. In other words, the legislation was designed to force foreign companies into choosing to do business with either Iran and Libya or the United States. ILSA, however, does not prohibit foreign companies from trading in Iranian crude oil and gas commodities.

Finally, in August 1997, President Clinton issued a third executive order that closed loopholes in the embargo whereby goods were being exported to Iran from third countries.[39] These executive orders and ILSA provide the general outlines of the Iranian trade restrictions. The U.S. Treasury Department's Office of Foreign Assets Control supplies the details and interpretations of the scope of the sanctions, issues regulations, and considers license applications.

The threat of secondary U.S. sanctions has deterred some multinational companies from investing in Iran. In August 1996, Australia's BHP withdrew from a proposed $3 billion pipeline project to transport Iranian natural gas to Pakistan and India under the threat of U.S. sanctions.[40] The Clinton administration's attempts to implement ILSA, however, have run into opposition from a number of foreign governments and companies. Following Conoco's abrogation of a contract to develop Iran's gas fields in

1995, the contract was awarded to a consortium composed of France's Total, Malaysia's Petronas, and Russia's Gazprom. In late 1996, the leaders of the European Union officially registered their opposition to the ILSA by passing Resolution 2271, directing EU members to not comply with the U.S. sanctions policy.[41] The economic tensions between the United States and the European Union escalated almost to the point of confrontation, so in May 1998 the Clinton administration decided to waive sanctions against Total.

This step was in line with ILSA, which allows the president to waive the mandatory sanctions. Under Section 4(c), the president may invoke a country waiver, exempting all nationals of a country from sanctions if their government has agreed to undertake substantial measures that would inhibit Iran's efforts to acquire weapons of mass destruction. No Section 4(c) waiver has ever been issued. Under Section 9(c), the president may waive sanctions upon a particular firm if it is important to the national interest. This waiver was intended to accommodate instances where invoking the sanctions could violate multilateral trade agreements, leading to unacceptable costs to U.S. economic interests or conflict with jurisdiction exercised by sovereign states.

Upon European insistence, the Clinton administration waived sanctions against the consortium led by Total based on Section 9(c), not Section 4(c). This was meant to be an exception, with other issues to be treated on a case-by-case basis, but it gave the impression that the U.S. government was not serious about penalizing IOCs investing in Iran. Indeed, although ILSA initially may have had some effect in deterring investment by companies that did not wish to risk sanctions, the law has never been enforced. Many European oil companies saw the waiver as a green light to do business with Iran with less worry about U.S. retaliation. Indeed, it can be argued that many European oil companies have benefited from the U.S. sanctions by not having to compete with American companies in developing Iranian hydrocarbon resources.

Sitting on the sideline, watching their European and Asian counterparts sign lucrative deals with the Iranian authorities, and at the same time being denied a chance to take part in these promising business opportunities by their own government, U.S. oil companies have increasingly voiced their criticism of the sanctions policy. They are more restrained by the executive orders than by ILSA (the former prohibit U.S. oil companies from investing in the Iranian energy sector while the latter focuses on foreign ones). It is important to point out that sanctions are programs imposed through executive orders and that regulations are created, amended, and terminated by the president and interpreted by executive agencies to reflect the administration's political objectives and perspectives at the time. Executive orders can be repealed in part or in whole without congressional approval. Leading U.S. oil companies, such as ExxonMobil, Chevron, and Conoco, have maintained contacts with the NIOC and top Iranian officials in the expectation that U.S. sanctions will be dropped or significantly relaxed in the near future. In late 2000 and early 2001, chief executives of these three U.S. oil giants met with Iran's Parliament Speaker, Mehdi Karroubi, and its foreign minister, Kamal Kharrazi, who were on separate visits to the United Nations in New York City. In addition, a British subsidiary of Conoco helped analyze seismic data collected in 1999 by the NIOC in its exploration of the Azadegan field, on the understanding that Conoco would be given priority in developing the field if and when it becomes legal to do so.

Given this constructive attitude of Conoco and other U.S. oil companies toward Iran, Iranian officials have called on Washington to get in line with U.S. oil companies and adopt a more favorable stance toward their country. Indeed, following the election of President Khatami in 1997, the Clinton administration took several steps aimed at reducing tension with Tehran. In July 1999, new regulations were issued to allow some exports of food, medicines, and medical equipment to Iran under specific conditions. (Still, export guarantees and trade financing were not permitted.) In April 2000, the ban on the import of caviar, pistachio nuts, and carpets was lifted. This move was largely symbolic, since these exports account for only a fraction of Iran's foreign exchange earnings, which come mainly from oil sales.

To sum up, by the early 2000s, U.S. sanctions against investing in Iran's hydrocarbon resources imposed since 1995 were still in place. It can be argued that at least initially these sanctions have slowed down foreign investment in Iran, but they have not succeeded in deterring many IOCs from signing expensive deals with Tehran. Put differently, it is likely that without U.S. sanctions Iran would have been more successful in attracting foreign investment to develop its oil and gas fields. The few steps President Clinton took before leaving office were largely symbolic and had no serious impact on the substance of either the bilateral relations between the two countries or the U.S. sanctions on Iran's energy industry. Under these circumstances, George W. Bush came to office in early 2001.

In the early months of the Bush administration, there were several signs that sanctions would be lifted and relations between Washington and Tehran would improve. This optimism was drawn from several statements made by top officials, particularly Vice President Dick Cheney and Secretary of State Colin Powell. Prior to his nomination, Cheney vigorously criticized sanctions as self-defeating and ineffectual in achieving their intended policy objectives. He also expressed the view that it was time to end the sanctions on U.S. energy companies because they gave their competitors in Europe and elsewhere a competitive advantage. Furthermore, Halliburton, the company that Cheney headed before returning to public life, has maintained some operations in Iran through foreign subsidiaries, which is all that is allowed under U.S. law. Similarly, in his confirmation hearings, Powell called Iran "an important country undergoing profound change from within."[42] He indicated that differences with Iran on policy should not preclude greater interaction, whether in more normal commerce or increased dialogue. Powell also told Congress that the administration was reviewing the United States' overall sanctions policy, with the aim of dismantling many of the sanctions regimes imposed in recent years.[43]

In spite of these statements and the early optimism, there was no relaxation of sanctions. Instead, two months after taking office, President Bush signed an executive order renewing the sanctions that were initiated by President Clinton in 1995. In August 2001, Bush signed into law the ILSA Extension Act of 2001, which provides for a five-year extension of the ILSA. Bush expressed serious concerns over Iran's support for terrorism, opposition to the Middle East peace process, and pursuit of weapons of mass destruction. Congress's stand has contributed to the renewal of sanctions. The Senate voted 96 to 2 to renew sanctions for five years.[44] In the House, 409 congressmen favored an extension, and 6 opposed it.[45] In other words, the great majority in both the Senate and the House voted against any relaxation of the sanctions. Given this overwhelming majority, it can be argued that the president decided not to spend any political capital on this issue.

Both the European Union and Japan strongly criticized the new law on the ground that it is an extraterritorial application of domestic law and a threat to the open international trading system. They vowed to take measures against the United States through the World Trade Organization if any action were taken against European or Japanese companies operating in Iran (and Libya). Understandably, Iranian officials expressed their frustration and anger at both the Bush administration and Congress.

To conclude, two concerns need to be addressed. First, there is a growing consensus in Washington that economic sanctions can advance national and global security objectives and can be an important foreign policy tool. Nevertheless, sanctions should be periodically reviewed to ensure their continued effectiveness and to minimize their costs on U.S. citizens and interests. Second, it is very unlikely that sanctions will be lifted or that there will be a fundamental shift in the relations between the United States and Iran unless the two countries reach an "understanding" on three issues: Tehran's opposition to the Arab-Israeli peace process, support to international terrorism, and attempts to acquire and develop WMD.

The Arab-Israeli Peace Process

The conflict between Israel and its Arab neighbors has been a major source of instability in the Middle East for decades. Despite attempts by several U.S. administrations to keep energy issues on a separate track from the Arab-Israeli conflict, developments in the Levant have always been echoed in the Persian Gulf. Major oil-producing countries like Saudi Arabia, Iraq, and Iran have always supported the Palestinian side against Israel and its main ally, the United States. Tehran's position is particularly interesting not only because Iran is not an Arab country but also because, unlike Riyadh and Baghdad, where the hostility to Israel has been consistent, Tehran's stand on the Arab-Israeli conflict has dramatically changed since the 1979 revolution.

Since its inception in 1948, Israel had enjoyed friendly relations with the Pahlavi regime. These close ties between the two countries were based on mutual economic and strategic interests. First, the shah had a profound admiration for Israel's technological advances, particularly in agriculture and water resources as well as in military technology. Second, with very limited hydrocarbon reserves, Israel was, and still is, heavily dependent on imported oil and natural gas.[46] Iran under the shah was a major oil supplier to Israel, particularly since 1957 when the Soviet Union terminated its oil shipments to Israel.[47] Third, Iran has the largest Jewish population in the Middle East outside Israel and was used by Israel as a transit point to facilitate the massive immigration of Aliyah from Iraq to Israel in the late 1940s. Fourth, some Iranian leaders believed that close relations with Israel would protect and promote their interests in the United States. Fifth, the two countries sought to resist and contain Soviet influence and communism in the Middle East. Sixth, radical Arab nationalism led by Egypt, Syria, and Iraq represented a threat to both Israel and Iran.

Taking all these economic and strategic concerns into account, one can understand the close cooperation between Iran under the shah and Israel.

Indeed, shortly after Israel was created, Prime Minister David Ben-Gurion articulated a strategy known as "Peripheral Alliance" in order to break the wall of isolation that surrounded it. The proposed alliance included Sunni Muslim but non-Arab Turkey, Shi'ia Iran, primarily Christian Ethiopia as well as the Christians of Lebanon, the Kurds in Iraq, and the non-Muslim population of Sudan. This alliance never materialized. Still, Tehran adopted a friendly and cooperative attitude toward Jerusalem and the Arab-Israeli conflict.

Unlike its Arab neighbors, Iran did not participate in any military confrontation against Israel. Indeed, Tehran granted Jerusalem de facto diplomatic recognition. Furthermore, in the aftermath of the 1973 Arab-Israeli war, the shah refused to take part in the oil embargo imposed by Arab producers and continued to sell oil to Israel at a normal price. Despite this close cooperation between Tehran and Jerusalem, the two sides strongly disagreed on the outcome of the 1967 Arab-Israeli war. The shah endorsed the United Nations Resolution 242, which called for an Israeli withdrawal from Arab land occupied during the war. Furthermore, the shah opposed Israeli control of Jerusalem. Despite these disagreements, relations between Iran and Israel remained strong until the collapse of the Pahlavi regime.

With the exception of Iran's relations with the United States, Iran's relations with Israel changed drastically with the outbreak of the 1979 revolution. In the early 1960s the shah stated, "Iran's relations with Israel are like true love that exists between two people outside of wedlock."[48] In contrast, Ayatollah Khomeini perceived Israel and Zionism as major enemies to Iran. Not surprisingly, Yasser Arafat was the first foreign dignitary to visit Iran after the revolution, and a Palestinian embassy replaced the Israeli diplomatic mission in Tehran. The visit was a recognition that the overthrow of the late shah, Israel's regional ally, and his replacement by a fiercely anti-Israeli regime represented a significant geopolitical shift in favor of the Palestine Liberation Organization. Relations between Iran and Arafat, however, rapidly deteriorated due to the Palestinian support to Iraq in its war against Iran (1980–88). When Arafat embarked on the peace process that led to the signing of the Oslo peace accords in 1993, he was denounced by Iranian leaders for surrendering to the "Zionist enemy." After Arafat signed the Wye River agreement in 1998, Ayatollah Khamenei, Iran's Supreme Leader, called Arafat a "traitor and a lackey of the Zionists."[49] Meanwhile, Iran gave substantial military and financial support to Hizbollah, Party of God, in its guerrilla war against Israel in southern Lebanon (1982–2000).[50]

Thus, since the 1979 revolution, most Israeli politicians have singled out Iran as the greatest threat to Israel and regional peace. Shimon Peres repeatedly stated that the Iranians saw the Israelis as "a collective Salman Rushdie."[51] Similarly, Ariel Sharon describes Iran as "the center of world terror."[52] On the other side, a significant faction within the Iranian religious/political establishment perceives no room for a compromise or a peaceful coexistence with Israel and calls for its annihilation. Still, another faction in Tehran has not ruled out the acceptance of Israel in case a peace agreement is reached with the Palestinians. Since 1998, President Khatami has reiterated that if the Palestinians accepted Israel's right to exist, Iranians would "respect the wishes of the Palestinian nation."[53]

Within this context (the conflicting signals coming out of Tehran), one can understand a potentially significant development in the Arab-Israeli conflict with dire implications on the Iranian role. In January 2002, the Israeli army seized a ship, Karine A, filled with more than 50 tons of arms bound for Yasser Arafat's Palestinian Authority in Gaza. The shipment included rockets, mortars, and antitank weapons, mines, explosives, rifles, and ammunition. These arms would have substantially increased the firing power available to the Palestinian Authority and is a clear violation of the Oslo Agreement. According to the Israeli government, the shipment was originated in Iran and Hizbollah, which has close ties with Iran, was also involved in the operation. This raises the possibility of a strategic alliance between Iran, Hizbollah, and the Palestinian Authority. Israeli diplomats explained the reasons behind Tehran's involvement. Iran wants a base and stronghold in the Palestinian Authority to threaten Israel. This is part of a larger effort to deter Israel from militarily trying to scuttle Iran's nuclear program.[54]

Iran denied any involvement with the weapons shipment. Instead, the government in Tehran said Israel was making the allegations to "intensify a crackdown on the Palestinian uprising."[55] Finally, some analysts suggested that the arms shipment might be the work of hardline elements inside the Iranian regime, aimed at stopping the apparent rapprochement with the West in the aftermath of the September 11 terrorist attacks and at embarrassing President Khatami and the reformers.[56] Given the highly classified intelligent information related to this operation, probably it will take years to know exactly who was involved and their motives. Still, this arms shipment represents a dangerous escalation of the proxy war between Iran and Israel.

To sum up, the Arab-Israeli peace process has almost completely collapsed and violence has intensified. The United States, the European

Union, and moderate Arab states with diplomatic relations with Israel (Egypt and Jordan) have been urging the Israelis and Palestinians to go back to the negotiation table and resume peace talks. At one point these efforts should succeed and the peace process will be revived. If the Israelis and Palestinians reach a peace agreement and implement it, how will Iran react? Rhetoric aside, there is an intense debate in Iran over the country's foreign policy, particularly relations with the United States and the Arab-Israeli conflict. In a symposium held in Tehran (December 2001) some academic experts and government officials argued that Iran does not need to be more extreme than the Palestinians themselves, who ostensibly are willing to make peace with Israel. They also underscored the damage caused to Iran's international status by its unqualified support for the Palestinians.[57] In response, Ephraim Halevy, the head of Mossad (the Israeli intelligence agency), said, "There is hope for moderation in the Iranian regime."[58] Still, given the division within the Iranian religious/political establishment, it is unlikely that Tehran would endorse efforts to make peace between Israel and the Palestinians in the foreseeable future. Rather, Iran might reluctantly stop opposing peace efforts.

Terrorism

In its annual report entitled *Patterns of Global Terrorism,* the State Department notes, "Iran remained the most active state sponsor of terrorism in 2000. It provided increasing support to numerous terrorist groups including the Lebanese Hizbollah, Hamas, and the Palestine Islamic Jihad, which seek to undermine the Middle East peace negotiations through the use of terrorism."[59] Similarly, James Woolsey, a former director of the Central Intelligence Agency (CIA), wrote, "U.S. policy is heavily driven by one decision that Tehran has made: to be the world's principal state-sponsor, encourager and bankroller of terrorism."[60] In short, Iranian involvement in terrorism has been a major concern for the United States since the 1979 revolution.

Shortly after the seizure of the American embassy in Tehran, diplomatic relations between the two countries were severed, and since 1984 Iran has been on the Department of State's list of states sponsoring terrorism. This means that Iran has been under various U.S. economic sanctions of increasing scope and intensity.

Most analysts agree that in the 1980s and early 1990s, officials at the highest level of the Iranian government were involved in planning and sponsoring terrorist attacks against both Iranian dissidents abroad and perceived enemies of the Islamic regime (American and Israeli targets). Since the election of President Khatami in 1997, however, Iranian sponsorship of terrorism has been endorsed by hard-line elements within the political/religious establishment in Tehran, while the moderate camp, led by President Khatami, has sought to condemn and distance itself from terrorism.

In a testimony before the Senate Select Committee on Intelligence, CIA director George Tenet said, "Khatami leads a group of moderates who are engaged in a genuine struggle with hard-line conservatives."[61] According to this line of argument, Iran is still involved in terrorism because President Khatami and his reformer supporters have not yet consolidated their control over the relevant security apparatus, particularly the Revolutionary Guard Corps and the Ministry of Intelligence and Security. Several cases can illustrate this lack of consensus among different factions within the Iranian religious/political establishment regarding the use of terrorism as a means of eliminating dissidents and achieving foreign policy goals.

Salman Rushdie

In 1989, the Ayatollah Khomeini issued a Fatwa (binding opinion) pronouncing a sentence of death on the Indian-born British author Salman Rushdie for writing a novel, *The Satanic Verses,* the content of which was considered to be blasphemous. Immediately the Fatwa was condemned by the ministers of foreign affairs of the members of the European Community, and senior-level diplomatic contacts between Brussels and Tehran were suspended. Then President Rafsanjani stated that the Fatwa was a religious issue, not a political one, implying that the government cannot change it, and declared that no official state organ would carry out the sentence against the British author. A diplomatic breakthrough came after contacts were stepped up following the elections of President Khatami in Tehran and the Labor government in London. Kamal Kharrazi, foreign minister of Iran, officially stated that his government would not take any action to threaten the life of Rushdie. In response, diplomatic relations at the ambassadorial level were restored between London and Tehran.

Mykonos Verdict

Another case is the so-called Mykonos Verdict. In 1997, the trial in Germany for the 1992 killing of Iranian Kurdish dissidents was concluded. The court stated that the government of Iran had followed a deliberate policy of liquidating the regime's opponents who lived outside Iran, including the opposition Kurdish Democratic Party of Iran. The judge further stated that the Mykonos murders had been approved at the most senior levels of the Iranian government by an extralegal committee whose members included the minister of intelligence and security, the foreign minister, the president, and the supreme leader. After the court ruling, all European Union countries except Greece recalled their ambassadors from Tehran. Furthermore, after a meeting of its foreign ministers in Luxembourg later in the month, the EU leaders adopted several measures to punish Iran, including halting bilateral ministerial visits and denying visas to Iranians holding intelligence and security posts. The election of Khatami in 1997, however, paved the way for a rapprochement between the two sides. In late 1997, the European ambassadors returned to Tehran, and the EU lifted the ban on ministerial contacts with Iran and called for expanded political ties.

Khobar Towers

In 1996, a bombing attack on Khobar Towers in Saudi Arabia killed nineteen U.S. servicemen. Three years later, evidence linked Iran to this terrorist attack. President Clinton sent a secret letter to President Khatami, asking for help in solving this case. This initiative did not produce any concrete results. Rather, in June 2001, indictments against thirteen Saudis and a Lebanese were issued. Attorney General John Ashcroft charged that Iranian officials "inspired, supported, and supervised those who carried out the bombing."[62] He also confirmed that the United States would be willing to pursue charges against Iranian officials if more evidence emerged.

Iranian officials categorically deny any association with terrorism. In an interview in November 2001, President Khatami said, "This is one of the injustices of the United States against us. Indeed, Iran is one of the biggest victims of terrorism."[63] The Iranian leader was referring to the Mujahidin-e Khalq' (MEK) attacks on targets inside Iran. The organization was formed in the 1960s by the college-educated children of Iranian merchants. Its philosophy is a mixture of Marxism and Islam. In the 1970s, the MEK killed several U.S. military personnel and civilians working on defense projects in Tehran and supported the takeover of the U.S. embassy in 1979. In the 1980s, however, the MEK fell out of favor with the clerical regime and was brutally destroyed. Massoud Rajavi, the organization's leader, and his followers fled to France and later to Iraq, where the MEK has been stationed since the mid-1980s. For almost two decades, the organization has been involved in vigorous attacks against the leaders of the Islamic Republic.

The Clinton administration and Congress have taken different approaches toward the MEK. In 1998, the State Department designated the organization under the Anti-terrorism and Effective Death Penalty Act of 1996, which makes it illegal for U.S. institutions and citizens to provide funds or other forms of material support to such groups. The law also makes members and representatives of those organizations ineligible for U.S. visas and subject to exclusion from the United States. On the other hand, in October 2000, 225 congressmen and 28 senators called for tough U.S. policies toward Iran and urged the State Department to remove the National Council of Resistance of Iran, the political wing of the MEK, from its list of terrorist organizations.

Meanwhile, Iranian leaders do not deny their close ties with Hizbollah (Party of God). This organization was formed in 1982 with Iranian help during Israel's invasion of Lebanon and began as a radical offshoot of Amal, a Shi'ia Muslim movement. The creation of Hizbollah and the consolidation of Iranian political and military presence in Lebanon were main achievements of Ali Akbar Mohtashemi, then Iranian ambassador to Syria. Hizbollah was involved in numerous anti-American terrorist attacks, including the suicide truck bombing of the U.S. embassy and Marine barracks in Beirut in October 1983 and the U.S. embassy annex in Beirut in September 1984. Members of the organization were responsible for the kidnapping and detention of U.S. and other Western hostages in Lebanon. This close association between Tehran on one side and Hizbollah and other militant organizations (e.g., Hamas and Jihad) on the other side has further complicated relations with Washington.

In 1996, Congress passed the anti-terrorism law, which permits U.S. citizens who are victims of terrorist acts abroad to sue foreign countries for civil damages in U.S. courts if those countries have been classified by the State Department as supporters of terrorism. In October 1998, Congress passed a measure that called for the State Department and the Treasury Department to assist victims of terrorism in locating money for judgments. But the legislation contained a waiver that permitted the president to decline support in the interests of national security.

Since then, federal judges have awarded significant payments to victims and families of alleged Iranian terrorism. Stephen M. Flatow, a New Jersey lawyer, was among the first to sue after his twenty-year-old daughter, Alisa, was killed in 1995 by a suicide bomber as she rode a bus in the Gaza Strip. Flatow contended that the Palestinian group Jihad was responsible for the attack and was funded primarily by Iran. In March 1998, a federal judge ordered Iran to pay $247.5 million. In a similar case in July 2000, a court awarded $327 million to the families of a New Jersey woman and a Connecticut man who were killed in a 1996 terrorist bombing in Israel. Three other Americans who were held in Lebanon in the late 1980s (Joseph J. Cicippio, Frank H. Reed, and David P. Jacobsen) won a $65 million judgment in federal court in 1998. Terry Anderson, a former Associated Press correspondent who was one of eighteen Americans taken hostages in Lebanon and who was held the longest (from 1985 to 1991), was awarded $324 million in 2000. In the same year, a federal judge ordered Iran to pay $355 million in damages to the family of Marine Lt. Col. William R. Higgins, who was taken captive and killed in 1989 while on a United Nations peacekeeping mission in Lebanon. A year later, June 2001, a federal judge ordered Iran to pay $323.5 million in damages to Thomas Sutherland, a university administrator who was taken hostage in Lebanon during the 1980s. Two months later, August 2001, a federal judge awarded the estate and family of the late Rev. Lawrence Jenco $314.6 million in damages from Iran for the eighteen months he was held hostage in Lebanon in the mid-1980s.

Thus, from 1998 to 2001, American federal courts ordered Iran to pay almost $2 billion to the victims of terrorism, claimed to be supported by Iran. Tehran has not responded to any of the lawsuits, losing all by default, and has shown no sign of paying the awards. Ironically, the Clinton administration opposed these payments. Officials in the State Department said that they wanted to help but maintained that much of the Iranian money was tied up in litigation before the Iran-U.S. Claims Tribunal under an agreement that led to the 1981 release of U.S. embassy employees who were taken hostage in Tehran. Administration officials also voiced concern that the sanctity of American diplomatic property abroad, a right extended to all diplomatic missions under international law, might be compromised if private citizens in the United States were allowed to seize foreign government property. In other words, there is a concern that countries might retaliate by seizing U.S. properties oversea. This deadlock between the Clinton administration and Congress was broken in October 2000 when the two sides agreed to put up the money with the expectation that the United States can get it back either through an international claims tribunal or through negotiations with Iran.

Under the Bush administration, the dilemma regarding financial compensation to the victims of Iran's involvement in terrorism has continued. Former employees at the American embassy in Iran who were held hostages for 444 days shortly after the revolution in 1979 filed a lawsuit against Iran demanding billions of dollars. The problem, however, is that in order to secure the hostages' release, President Carter signed an agreement known as the Algiers Accords. The terms of the agreement required the U.S. government to "bar and preclude" lawsuits by the former hostages. In response, in November 2001, Congress passed and President Bush signed an amendment to the Anti-Terrorism Law that gave the courts specific jurisdiction over the Tehran hostages' case. Thus the question of Tehran's involvement in terrorism will have to be appropriately addressed in order to have a realistic chance for a rapprochement between the United States and Iran. This conclusion can be equally applied to another stumbling block—Tehran's efforts to acquire and develop nonconventional weapons.

Weapons of Mass Destruction

Since the end of the Cold War, the United States has identified the proliferation of weapons of mass destruction—nuclear, biological, and chemical (NBC) weapons and the missiles that carry them—as one of the most significant threats to its national security. A recent study published by the U.S. Defense Department describes Iran as "one of the countries most active in seeking to acquire NBC and the missile-related technologies," adding that "Iran's efforts continued in the last several years notwithstanding President Khatami's moderation of the regime's anti-Western rhetoric."[64] Iran categorically denies any interest in acquiring or developing these weapons.

Still, scholars and policymakers have identified several incentives for Iran to develop such capability. First, WMD, particularly nuclear weapons, can increase Iran's influence and prestige throughout the Islamic world. Second, WMD can limit U.S. influence and presence in the Middle East, especially in the Persian Gulf. These nonconventional weapons can contribute to diminishing the gap in military capability between Tehran and Washington. Third, Israel, India, and Pakistan have nuclear devices, and Egypt and Syria are believed to possess chemical and biological weapons and missiles. Several of Iran's neighbors, including Saudi Arabia, have missiles that can reach Iran. So it is quite natural that Iran would want a similar capability. Fourth, the main reason for Tehran's interest in WMD is its experience during the war with Iraq (1980–88). Iran was no match for Iraq's chemical and missile capabilities. This gap was a major reason for the collapse of the Iranian army at the end of the war. As one scholar argues, "Arguments for an Iranian WMD capability boil down to the need to counter an uncertain Iraq."[65] Indeed, given the territorial disputes between the two countries and the fact that most Iraqis, who are Shiʿia, are ruled by the Sunni minority, suspicion will continue to characterize the relations between the two countries. In short, regardless of who rules in Baghdad, Iraq will always be an important factor in Iran's security configuration.

Many Iranian leaders see the possession of long-range missiles as vital to national security. Given that Iran lacks free access to the most sophisticated aircraft and the necessary spare parts, missiles can be seen as an attractive option. Missiles are relatively cheap and easy to manufacture domestically. They are easy to conceal, certain to penetrate, and do not require trained pilots.[66] Furthermore, there is no international treaty to which Iran is a party that limits its capacity to develop missiles. Consequently, Iran has invested substantial resources to produce indigenous missiles. Most prominent are Shahab-3 and Shahab-4, which can reach almost any target in the Middle East and even in Europe.

Iranians argue with great emotion that they, more than any other country, have an interest in seeing the Chemical Weapons Convention (CWC) succeed, since they were subject to chemical attacks by Iraq during the war. Indeed, Tehran has signed and ratified the CWC and admitted developing a chemical warfare program during the 1980s as a deterrent against Iraq. Moreover, Iran claims that after the 1988 cease-fire, it terminated its program. In 1999, a team from the CWC visited Iran and confirmed the accuracy of its claims.[67] Nevertheless, the U.S. government believes that Iran is continuing to seek production technology, expertise, and precursor chemicals from entities in Russia and China that could be used to create a more advanced and self-sufficient chemical warfare infrastructure.

In 1972, 143 countries, including Iran, signed the Biological Weapons Convention, which bans the production and use of biological weapons. This treaty, however, has no verification mechanism. Between 1993 and 2001, an ad hoc group, the Geneva-based Conference on Disarmament, worked toward a draft verification protocol. These efforts collapsed when the Bush administration decided not to sign a new agreement. In other words, like all other countries, Iran claims that it is in compliance with the Biological Weapons Convension, but there is no mechanism to verify its claims. Washington, on the other hand, believes that Tehran has a growing biotechnology industry, significant pharmaceutical experience, and the overall infrastructure to support its biological warfare program. The dual-use nature of the materials and equipment being sought by Iran further complicates an accurate assessment of the country's current and potential biological capabilities.

Unlike Israel, India, and Pakistan, Iran signed the Non-Proliferation Treaty, and since the early 1970s Iran has called for the creation of a nuclear-free zone in the Middle East. According to the terms of the treaty, the five nuclear powers—the United States, Britain, France, Russia, and China—are obligated to share their nuclear expertise and technology with other members in the treaty who pledged not to develop nuclear weapons. The reason is to help them benefit from the civilian use of nuclear energy. Thus Iran has the legal right, under international law, to receive help in nuclear technology from any of these five nuclear powers. The United States, however, strongly opposes any cooperation with Iran on the grounds that Tehran would transfer the nuclear knowledge and materials it receives from civilian uses to military ones. In order to prevent Iran from receiving nuclear materials, the United States has sought cooperation from other nuclear powers, particularly China and Russia.

In 1997, Beijing promised Washington that it would suspend any nuclear cooperation with Tehran. American officials are convinced that China has lived up to its commitments. In 1995, under heavy pressure from the Clinton administration, Russian president Boris Yeltsin promised that Russia would not provide Iran with uranium enrichment technology of any kind, although it would go ahead with a contract to complete a civilian nuclear reactor at Bushehr. In spite of this promise, U.S. officials believe that nuclear materials from Russia have been smuggled into Iran. In response, in 1999 the United States placed sanctions on seven Russian companies and three institutions accused of aiding Iran's nuclear efforts.

One way to end this deadlock between Washington and Tehran over nuclear weapons is for the latter to ratify two additional protocols to the International Atomic Energy Agency. These two protocols, sometimes known as Program 93+2, are designed to avoid the disastrous experience with Iraq.[68] Iran has refused to ratify the 93+2 provisions on the grounds that it is still being denied civilian nuclear technology.

In closing, the question of what will happen next between the United States and Iran in the area of WMD proliferation needs to be addressed. Five conclusions can be drawn from the foregoing discussion and are likely to continue to shape these accusations and counteraccusations between the two nations. First, for the foreseeable future, Iran will continue to be monitored very closely for any clandestine activity to acquire and develop nonconventional weapons. Second, Iran, backed by most Arab states, will continue to argue that there is a double standard in U.S. policy concerning the nonproliferation regime. Indeed, in any debate about nuclear weapons, one of the weakest points in the U.S. argument has been how to avoid condemning Israel while focusing on Iran and other Arab countries.

Third, Iran's reasons for acquiring and developing WMD have less to do with the nature of the political regime in Tehran and more to do with a perception of fundamental threat to its national security. In other words, a major explanation for Iranian interest in NBC weapons has been to deter its traditional enemy, Iraq. This has been a constant factor in Tehran's security configuration under both the shah and the Islamic regime. Iran ruled by a moderate faction or a hard-liner group will likely have the same perception and pursue the same policy. As one analyst wrote, "Even if Thomas Jefferson became president in Tehran, Iran will continue to pursue the bomb."[69] Fourth, in its attempt to neutralize Iranian efforts to acquire nonconventional military capabilities, the United States has relied on a variety of approaches including export controls, sanctions, and pressuring allies and adversaries to follow suit. However, the preferred (and most difficult) approach to addressing Iran's WMD question is within the context of a regional security framework that would follow a resolution of major regional conflicts (the Arab-Israeli dispute and stability in the Persian Gulf). Finally, the evolution of a prosperous market economy, civil society, and competitive and accountable political system inside Iran is likely to lead to greater openness and moderation in its foreign policy and reduce the need for WMD. The United States should encourage this trend.

Despite these critical and unresolved issues between Washington and Tehran (Iran's opposition to the Arab/Israeli peace process, sponsorship of terrorism, and attempt to acquire and develop WMD), the September 11 terrorist attacks and the war to destroy the Taliban regime in Afghanistan have allowed the two countries to work together and cross the gulf of mistrust that separates them.

Iran and the War on Terrorism

For centuries, Iran has had strong cultural, economic, and political ties with Afghanistan. The two countries share a 560-mile border. Furthermore, about 20 percent of Afghans are Shi'ia (90 percent of Iranians are Shi'ia). In addition, until the communist seizure of power in 1978, Persian was the language of the court and educated elite. No wonder Iran was a dominant power in Afghanistan for centuries. When the pro-Soviet communist regime fell in 1992, a new government loyal to Iran took over, the remnants of which later formed the Northern Alliance, the main opposition group that fought the Taliban. But Pakistan, with generous Saudi funding, then organized a new movement, the Taliban, which in 1996 drove Iran's allies from Kabul. After the Taliban seized power, Iran hosted and supported the main Afghani opposition leaders, including Burhanuddin Rabbani, Ahmad Shah Masood, and Ismail Khan.

As a fundamentalist Sunni movement, the Taliban considered Shi'ia Iran to be an affront to Islam and a mortal enemy. In return, Iran and Russia provided weapons and training to the anti-Taliban Northern Alliance (previously known as the United Front), particularly to Afghanistan's mostly Shi'ia Hazara minority and Persian-speaking Tajiks. In 1998, the Taliban forces overran the northern city of Mazar-e Sharif and killed eight Iranian diplomats. In response, a war almost erupted between the two sides. The Taliban ruled most of Afghanistan from 1996 until 2001. Under their rule the Iranian/Afghani relations were characterized by mutual hostility and suspicion. After the September 11 terrorist attacks, the United States decided to retaliate and eliminate the al-Qaeda organization and its main ally, the Taliban. Tehran had many reasons to support this war:

(1) The removal of the Taliban would provide new dynamics under which a less hostile and friendlier government in Kabul might be established. Put differently, Iran would like to see a friendly government right on its eastern border.
(2) Political stability in Afghanistan would facilitate the return of more than 2 million Afghani refugees in Iran. Given the economic hardship facing many Iranians, including unemployment and inflation, Tehran had tremendous difficulty coping with this huge number of Afghani refugees.
(3) A strong central authority in Kabul would likely stop or reduce the production of opium and other illegal drugs. In the late 1990s, thousands of Iranians were killed fighting drug traffickers along the border with Afghanistan.

Given these incentives, Tehran took several steps to cooperate with the United States in the war on al-Qaeda and the Taliban. These include both symbolic and concrete actions. A few days after September 11, Tehran mayor Morteza Alviri sent a message of condolence to New York City mayor Rudolph Giuliani in the first public official contact between the two countries since the 1979 revolution. President Khatami strongly condemned what he called "the anti-human and anti-Islamic acts" and, in a letter to UN secretary-general Kofi Annan, suggested that a meeting of heads of states be held to adopt measures to fight terrorism. Meanwhile, Iran agreed to rescue any American troops in distress in its territory and provided a port for shipping American wheat into the war zone in Afghanistan. Equally important, Iranian diplomats played a constructive role behind the scenes in the Bonn negotiations that led to the formation of a coalition government in Kabul in late 2001.

These Iranian measures to endorse the American-led war on al-Qaeda and the Taliban and the attempts by some moderates within the religious/political establishment in Tehran to mend fences with Washington were strongly restrained by the conservative elements in the Iranian leadership. The supreme leader, Ayatollah Khamenei, warned that the United States was not "fit" to lead an international antiterrorist effort.[70] Khamenei also stated Iran had nothing to gain and everything to lose from official contact with the United States. Former president Rafsanjani argued along the same lines: "If the Americans think they can intensify the pressure on us, Lebanon and the Palestinians, by penetrating Central Asia, the Caucasus, and by completing the process of besieging Iran, they should know that the crisis will expand in the long run."[71] In other words, top leaders in Iran believe that the United States is using the terrorism issue as a pretext to establish a long-term presence in the region, not to combat terrorism but to enhance and protect U.S. commercial interests, encircle Iran with U.S. allies, and impose U.S. policies in the area through force. They accuse Washington of seeking to define terrorism to suit only its own ends, and they would like the war on terror to be led by the United Nations, not the United States.

On the other hand, shortly after eliminating the Taliban regime and establishing a new government in Kabul, top U.S. officials expressed suspicion about Iran's intentions and policies. According to the U.S. government, in late 2001 and early 2002 Iran gave safe haven to small numbers of al-Qaeda fighters fleeing Afghanistan and supplied arms, money, and training to the tribal leaders and militia commanders who run the western Afghan city of Heart and surrounding areas. The goals are to destabilize the government in Kabul, weaken Western influence, and consolidate Iran's own power. President Bush warned the Iranians, "If they in any way, shape, or form try to destabilize the government in Kabul, we will deal with them."[72] Iranian officials categorically denied these accusations.

It will take some time to assess the long-term impact of the Iranian stand in the American war against al-Qaeda and the Taliban on the relations between Tehran and Washington. Still, some tentative conclusions can be drawn. First, the war in Afghanistan and its aftermath show that Tehran and Washington had similar but not identical goals regarding the Taliban. The Taliban were a common enemy to both the United States and Iran, and they were destroyed. But the story does not end there. The destruction of the Taliban paved the way to the establishment of a new government led by Hamid Karazi that enjoys close ties to the United States. Furthermore, in the aftermath of the war, American military presence and bases are in several Central Asian states that are close to Iran's eastern borders (e.g., Uzbekistan and Pakistan). On the other side is Turkey, a NATO member that hosts U.S. military bases. Furthermore, Washington has military facilities and bases in the Arab states on the Persian Gulf. Tehran finds itself surrounded by American military almost in all directions. Given the relations between Washington and Tehran for more than twenty years, many Iranian leaders see the heavy American military presence as a threat to their country.

Second, if calm and stability return to Afghanistan, interest in pipelines to carry Caspian Sea oil and gas across the country to Pakistan and beyond may revive, undercutting Iran's ambition for such pipelines to take cheaper, shorter routes across its territory and depriving it of lucrative source of revenues and political leverage. Third, because of these current and potential strategic and economic risks, Tehran is determined not to be left out of Afghanistan. Before September 11, Iran participated in the UN negotiating process known as "6+2," encompassing the six neighbors of Afghanistan (China, Pakistan, Iran, Turkmenistan, Uzbekistan, and Tajikistan) and two external powers (Russia and the United States). Since September 11, Tehran has been reasserting its influence through the cultural and historical ties it has with Afghanistan. Probably Iran cannot determine the outcome in Afghanistan. But, as in the Persian Gulf and Lebanon, Tehran should be included in a final solution.

Fourth, it looks as if the Iranian stand in the war against terrorism has enhanced the country's international image, contributed to better relations with Europe, and reduced tension with the United States. British foreign secretary Jack Straw visited Tehran twice in one month shortly after September 11. This was the first visit by a British foreign secretary to Iran since the 1979 revolution. Another symbolic gesture was a handshake between Secretary of State Colin Powell and his Iranian counterpart, Kamal Karrazi. Finally, several members of Congress invited Iran's UN ambassador to a dinner at the Capitol and discussed ways to ease more than two decades of estrangement.[73]

What Lies Ahead?

For more than twenty years, the relations between Washington and Tehran have been characterized by suspicion, antagonism, and open hostility. The Ayatollah Khomeini used to say that the revolution did not take place because the prices of watermelon were high. By this he meant that the revolt against the shah's regime was not incited only by economic reasons. Rather, the revolution was justified by the widespread belief that the shah had allowed the United States to corrupt Iranian society. Accordingly, standing up against the United States was (and for some still is) a fundamental goal for the Islamic Republic. Indeed, confronting the United States partially legitimized the clerical regime.

On the other side, the United States has perceived the new regime in Tehran as a menace trying to destabilize the Middle East by opposing the Arab-Israeli peace process, sponsoring terrorism, and stockpiling WMD. In order to undermine Iran's capability to pursue these goals, the United States declared what can be described as economic warfare against Iran. In addition to imposing economic sanctions, Washington has sought to use its political and financial leverage in international institutions to isolate Iran and deprive it of badly needed financial resources. The World Bank and the World Trade Organization are cases in point.

Iran was an active borrower from the World Bank until 1976. After a devastating earthquake in 1990, Iran received six loans totaling $843 million. However, under pressure from the Clinton administration, the bank turned down requests for more money. In 2000, the United States failed to block approval of two loans for Iran: $87 million for health care and nutrition and $145 million for the Tehran sewage project.[74] Since then, Iran has applied and received more loans. In approving loans to Iran, the World Bank's board of directors expressed support for the reform agenda of President Khatami. Similarly, for several years Iran has sought to join the WTO. Officially, Iran submitted an application to join the WTO in 1996. Tehran has already taken several steps to strengthen its application. These include enacting a foreign investment law, reforming its tariff structure, and recognizing and enforcing international arbitration awards, both of which are basic requirements for WTO membership.[75] Furthermore, several Arab and Muslim countries as well as members of the EU support Iran's application. Nevertheless, under the WTO's consensus system for reaching decisions, the United States has been able to block Iran's request to negotiate with the international body.

Economic and political developments in Iran and the United States since 1979 have underscored several new perceptions and realities that are likely to shape future relations between the two nations. On the Iranian side, hostility to the United States is seen less as a source of internal legitimacy than it was in the early 1980s. Instead, improving economic conditions and the standard of living is more crucial. Lifting American sanctions would further encourage foreign investment in Iran, facilitating and accelerating the country's reintegration into the international economic system. Clearly, the rest of the world has adopted a different approach toward Iran than the one chosen by the United States. Consequently, Washington has failed in its attempt to isolate Tehran. Furthermore, U.S. corporations have lost lucrative business opportunities for their Canadian, European, and Asian competitors.

Relations between Washington and Tehran do not need to be seen in zero-sum terms. There are several areas where the two nations have mutual interests. Containing Saddam Hussein and restoring stability in Afghanistan are good examples. Most important, Iran has always been a major player in the global energy market. With its massive oil and natural gas reserves and its strategic location on the Persian Gulf and the Caspian Sea, Tehran has a significant role to play in promoting stability in the two regions and helping meet future world energy needs. Indeed, Iran can be seen as an energy giant with one foot in the Persian Gulf and the other in the Caspian Sea.

6

The Geopolitics of the Caspian Sea

The Caspian Sea is located in northwest Asia. Azerbaijan, Iran, Kazakhstan, Russia, and Turkmenistan share the Caspian Basin. Their policies on the exploration and development of the region's hydrocarbon resources since the collapse of the Soviet Union in late 1991 are the focus of this chapter. The region is important to the United States and other energy-consuming countries because it can contribute significantly to the world's oil and gas production and to the diversification of global hydrocarbon resources and reduce dependence on the Persian Gulf. In short, the Caspian Sea has the potential to substantially enhance global energy security.

The region is not new to the petroleum and natural gas industry. Commercial energy output began in the Caspian Basin in the mid-nineteenth century, making it one of the world's first energy provinces.[1] By 1900, the Baku region produced half of the world's total crude oil. This impressive level of production was the result of combined efforts and investment by the Nobel brothers, the Rothschilds, and the leaders of Royal Dutch/Shell, who helped Russia to develop Caspian oil resources.[2] This oil carried considerable strategic weight in both world wars. Short of fuel, the German army sought unsuccessfully to capture the Baku region. Germany's failure to secure access to the Caspian's oil resources was a major reason for its defeat in 1918 and 1945. Indeed, some of the most brutal battles during the Second World War were fought north of Azerbaijan.

Since the early 1950s, however, several developments contributed to a substantial reduction of Caspian oil production. Concern over Baku's vulnerability to attack during World War II, along with the discovery of oil in the Volga-Urals region of Russia, and later in western Siberia, led to a switch in the Soviet Union's investment priorities. This new policy resulted in decreased exploration and production in the Caspian for most of the second half of the twentieth century. Since the late 1980s, however, Azerbaijan, Kazakhstan, and Turkmenistan have gradually occupied a central

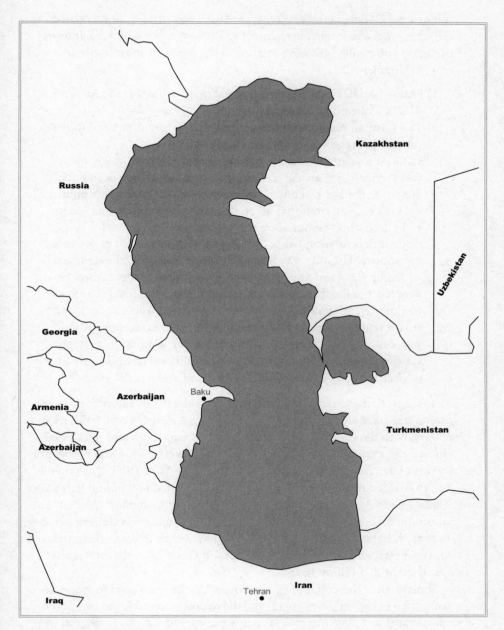

Map 6. The Caspian Sea

stage in the global energy markets. The three countries have succeeded in attracting massive foreign investment to their oil and gas sectors. International oil companies' growing interest in the region is in response to several developments:

(1) Production has declined in such great oil provinces as the Alaskan North Slope and the North Sea.
(2) The Caspian region contains some of the largest underdeveloped oil and gas reserves in the world.
(3) Saudi Arabia and Kuwait were strongly opposed to foreign investment in their energy sector. Iran, Iraq, and Libya have been under U.S. unilateral or UN multilateral economic sanctions. For political reasons, international oil companies have been denied access to hydrocarbon resources in the Middle East.
(4) Upon independence, Baku, Astana, and Ashgabat had very weak economies. The only way to stop the deterioration of the standard of living was (and still is) to fully utilize their hydrocarbon resources. But the three states lack the necessary financial resources to explore and develop oil and gas fields. Their leaders concluded that foreign investment was essential for their economic growth. Indeed, the International Energy Agency estimates that the Caspian region needs approximately $200 billion in investment in order to fully develop and utilize its oil and gas resources.[3]

Taking all these developments into consideration, international oil companies have negotiated and signed several deals, worth billions of dollars, with the Caspian states, particularly Kazakhstan and Azerbaijan. One of the earliest (and largest) deals was a $20 billion joint venture between Chevron and Kazakhstan, concluded in April 1993. The scheme was to develop the Tengiz oil field, with an estimated 6–9 billion barrels.[4] Similarly, the Azerbaijan International Operating Company (AIOC), an international consortium, signed an $8 billion production sharing agreement with Baku in September 1994. They agreed to develop three fields: Azeri, Chirag, and the deepwater portions of Gunashli, with total reserves estimated at 3–5 billion barrels.[5]

Shortly after the collapse of the Soviet Union, the Caspian states became open to foreign investment, and the region has reemerged as a potentially significant player in global energy policy. Indeed, the geological potential of the Caspian region as a major source of oil and gas is not in doubt. The rate of investment, however, is determined by the perceived risk in the region or what industry experts call "above-the-ground risk."

In other words, the risk is not in finding the oil and gas but in juggling the multitude of risks associated with operating in very difficult host country environments. These challenges include an accurate assessment of the Caspian region's hydrocarbon resources; rivalries between regional and international powers; domestic ethnic conflicts; disputes over the legal status of the Caspian; and disagreements over the most cost-effective pipelines.

Since the early 1990s, these five challenges have slowed down and restrained the exploration and development of Caspian states' mineral resources and raised doubt on the potential impact the region might have on global energy markets. Indeed, they underscore the fact that the exploitation of hydrocarbon resources has always been as much about politics as markets. This chapter examines these five real and potential challenges. It argues that the region could help increase world energy security by diversifying global sources of supply, although the notion that the Caspian's oil and gas can be the panacea to long-term global energy security is misguided. The region's potential impact on global petroleum policy should be neither overestimated nor underestimated.

Oil and Gas Resources: An Assessment

The intense interest shown by international oil and gas companies in the Caspian Sea testifies to its promising potential. Indeed, some top officials in Kazakhstan and Azerbaijan describe their countries as "another Middle East," "another Saudi Arabia," and "another Kuwait." This euphoria, concerning an energy bonanza from the region, raises the question of how much oil and natural gas the Caspian Sea has. Ironically, there is a great disparity between the assessments provided by the U.S. government and those provided by academic institutions and industry experts. Tables 6.1–6.3 illustrate some of these disagreements.

According to some analysts, this inconsistency in the assessment of the Caspian region's oil and gas resources reflects an orchestrated effort by Washington to exaggerate the significance of the region's hydrocarbon wealth. The goal is to reduce American and Western energy dependence on the Persian Gulf. For example, the Petroleum Finance Company claims that the U.S. government has published inflated and unsubstantiated numbers.[6] Similarly, a study by the International Institute for Strategic Studies states that "despite their near universal quotation, the U.S. Department of Energy figures are generally perceived to be an order of magnitude away from reality."[7] The Oxford Institute for Energy Studies concludes, "A

Table 6.1. British Petroleum/Amoco's estimate of the Caspian's resources, 2000

Country	Oil reserves (billion barrels)	Natural gas reserves (trillion cubic feet, tcf)
Azerbaijan	6.9	30.0
Kazakhstan	8.0	65.0
Turkmenistan	0.5	101.0
Total	15.4	196.0

Source: BP/Amoco, *BP Statistical Review of World Energy* (London, June 2001), 4, 20.

Table 6.2. *Oil and Gas Journal*'s estimate of the Caspian's resources, 2000

Country	Oil reserves (billion barrels)	Natural gas reserves (trillion cubic feet, tcf)
Azerbaijan	1.2	4.4
Kazakhstan	5.4	65.0
Turkmenistan	0.5	101.0
Total	7.1	170.4

Source: "Worldwide Look at Reserves and Production," *Oil and Gas Journal* 98, no. 51 (December 18, 2000), 122.

Table 6.3. U.S. Department of Energy's estimate of the Caspian's resources, 2000

Country	Oil reserves (billion barrels)			Gas reserves (tcf)		
	Proven	Possible	Total	Proven	Possible	Total
Azerbaijan	3.6–12.5	32	36–45	11	35	46
Kazakhstan	10.0–17.6	92	102–110	53–83	88	141–171
Turkmenistan	1.7	80	82	98–155	159	257–314
Total	15.3–31.8	204	220–237	162–249	282	444–531

Source: Energy Information Administration, *Caspian Tables and Maps* (July 12, 2000), on-line at www.eia.doe.gov.

better point of comparison is the North Sea rather than Saudi Arabia. The possibility of the Caspian serving as a major long-term competitor and substitute for the Gulf then evaporates."[8]

In addition to these divergences, anyone attempting to assess Caspian oil and gas resources should consider several factors. First, since the early 1990s, most of the efforts by the IOCs have focused on the three former Soviet republics of Azerbaijan, Kazakhstan, and Turkmenistan. The other two Caspian littoral states—Iran and Russia—are believed to have little, if any, oil and gas resources in their shares of the sea. However, since the late 1990s, both Tehran and Moscow have sought to challenge this notion. In 1999, Lasmo and the Royal Dutch/Shell Group began a seismic survey for oil exploration off Iran's Caspian coast.[9] The preliminary results of this survey showed that there are approximately 10 billion barrels of in-place crude of which 3 billion barrels are recoverable.[10] Similarly, in March 2000, Russia's Lukoil announced that it had identified six structures holding more than 2 billion barrels of sweet, low-sulfur crude in the Russian offshore section of the Caspian.[11] There are signs that both Iran and Russia have some oil and gas deposits in their respective shares of the Caspian Sea. The exact quantity of these deposits is yet to be fully assessed.

Second, in spite of the euphoria over the potential of Azerbaijan, Kazakhstan, and Turkmenistan, drilling has been disappointing. For instance, in July 2001, ExxonMobil suspended all operations in the Oguz field off Azerbaijan because the well it had drilled turned out to be dry. In the same month, Chevron stopped work on the Absheron field off Baku because only a very thin layer of gas was discovered there.[12] These disappointing results do not mean that IOCs will lose interest in the region. Rather, they call for a cautious approach in assessing the Caspian's oil and gas potentials.

Third, in spite of some disappointments, there have been some exciting discoveries in offshore Azerbaijan and Kazakhstan, particularly Shah Deniz and Kashagan. In July 1999, BP/Amoco announced it had struck natural gas at Shah Deniz, one of the major offshore hydrocarbon discoveries in the Caspian Sea since the breakup of the Soviet Union. The field, which is thought to be the world's largest gas discovery since 1978, is estimated to contain between 25 and 39 tcf of natural gas.[13] It is being developed by an international consortium of companies from the United Kingdom, Norway, Turkey, Russia, France, Azerbaijan, and Iran. It is important to point out that from the start Shah Deniz was planned as a largely European venture—without any U.S. participation—so that room could be found to accommodate an Iranian shareholder. This was necessary because when Azerbaijan was negotiating the development of the original Azeri-Chirag-Gunashli concession with members of what was to become the AIOC, it promised Tehran a 10 percent stake in the venture. Under strong U.S. pressure, Baku reneged on this commitment and agreed to grant the Iranians a stake in Shah Deniz.

Seismic studies and preliminary drilling in the offshore Kashagan block by the Offshore Kazakhstan International Operating Company (OKIOC)[14] have recently turned up spectacular results, with analysts estimating possible oil reserves of up to 40 billion barrels of which 10 billion barrels are thought to be recoverable.[15] This suggests that Kashagan has the potential to produce up to 2 million barrels of light crude a day once the field is fully developed. Accordingly, the field is considered one of the largest oil discoveries in the world in the last several decades. However, Kashagan bristles with so many technical problems that Soviet energy officials, who became aware of it in the 1980s, set it aside to concentrate on Siberia. The challenges include ice-filled shallow water, high gas and sulfur content in the oil, the difficulties of getting heavy equipment to an inland sea, and the need to tread lightly in a nature preserve.[16] Still, these two significant discoveries—Shah Deniz and Kashagan—have triggered upward revisions of the Caspian oil and gas potentials and added a sense of optimism that the Caspian region can play a vital role in meeting world energy demand.

To sum up, the Caspian Sea's hydrocarbon resources have yet to be fully assessed. Since the early 1990s, some significant discoveries have been made and some dry wells have been found. Still, much of the region's resources remain to be explored and developed. It is likely that the Caspian's oil and gas reserves will continue to be revised upward. Meanwhile, costs of exploration and development are projected to decline as the drilling and technological infrastructure develops. Many energy experts predict that the Caspian will play a significant role in world energy markets. Wood Mackenzie, a leading world oil consultant, suggests that the Caspian region has the potential to produce around 3.8 million barrels per day (b/d) by 2015.[17] Most of this oil will come from Kazakhstan and Azerbaijan. Turkmenistan's fortune is more concentrated in its natural gas. This projection and similar ones by other industry experts underscore three important conclusions:

(1) The Caspian Sea oil and gas potential are not in doubt.
(2) Still, a realistic assessment suggests that the Caspian region is less likely to become "another Middle East" and more likely to resemble the North Sea.
(3) The full and speedy utilization of the region's hydrocarbon resources depends on addressing geopolitical differences between regional and international powers, solving domestic ethnic conflicts, determining the legal status of the Caspian, and constructing a variety of pipelines.

Geopolitical Considerations

Russian/Soviet domination of Central Asia and the Transcaucasus region lasted for more than a century. The breakup of the Soviet Union in December 1991 created a "power vacuum" and triggered jockeying for influence by both neighboring states and a distant superpower. Some analysts describe the geopolitical rivalry in the region as a "neo–Cold War" or "neo–Great Game."[18] Indeed, the creation of several regional organizations testifies to this perception of power vacuum and the need for regional economic and security cooperation. These new regional organizations include the Commonwealth of Independent States (CIS),[19] Partnership for Peace,[20] Black Sea Economic Cooperation Pact,[21] the Shanghai Forum,[22] GUUAM,[23] and Economic Cooperation Organization.[24] Most of these organizations were created, or expanded, after the collapse of the Soviet Union. Some of them compete with each other, and some have similar strategic goals. The experience of regional and international rivalry in Central Asia and Transcaucasus since the early 1990s suggests several factors that need to be considered when analyzing the region. First, militarily, Central Asia and Transcaucasus states are surrounded by four nations with nuclear capability: Russia, China, India, and Pakistan. Turkey is a NATO member, and Iran has had hostile relations with the United States since 1979. Furthermore, instability in Afghanistan that led to the September 11, 2001, terrorist attacks and the subsequent war has been a constant challenge to the overall regional stability. In short, the military and strategic balance of power in the region is very fragile. Second, the Caspian's hydrocarbon wealth is at the heart of this rivalry between regional and international powers. Every contender wants a piece of the potentially large energy pie in the Caspian. Third, ironically, development of the region's energy resources has been slowed by the uncertainty of fiery competition between regional and international powers. Fourth, most contenders came to realize that their rivalry over the Caspian's energy resources should not be put in zero-sum terms. Rather, the Russians, Americans, Iranians, and Turks have found common ground and are working together on several schemes to develop the region's oil and gas fields and pipelines. Fifth, there is an internal competition between strategic and economic interests within each country. In other words, there is no monolithic American or Russian policy toward the Caspian. Instead, foreign policy goals and strategic considerations compete with commercial ones in formulating and determining where each of these players stands.

Taking all these factors into consideration, the analysis briefly examines the main guidelines of American, Russian, Iranian, and Turkish policies in the Caspian region since the early 1990s as well as the potential impact of the war in Afghanistan on Central Asia's energy map.

The United States

The collapse of the Soviet Union in 1991 presented the United States with a great opportunity to enhance its commercial interests and energy security. The Caspian region, which was closed to foreign investment during the Soviet era, became open to IOCs. Eager to underscore their economic independence from Moscow and develop their main source of foreign currency (oil and gas revenues), the leaders of Azerbaijan, Kazakhstan, and Turkmenistan invited and encouraged foreign companies to invest in developing their energy resources.

Given that the United States is the world's largest oil consumer and importer, the opening of the Caspian states to foreign investment represented a great opportunity that Washington does not intend to miss. Washington has several objectives in the Caspian region. First, it supports the process of nation-building within each of the newly independent states and strengthening their economic and political independence from Moscow. This includes attempting to solve ethnic conflicts and promoting regional stability. Second, the exploration and development of the Caspian's hydrocarbon resources would increase the global oil and gas supplies and contribute to Washington's goal of energy security by diversifying suppliers. In other words, the development of the Caspian energy resources would reduce U.S. and world dependence on the Persian Gulf. Third, the United States opposes Russian monopoly over oil and gas pipelines from the Caspian Sea to the international markets. More routes mean less threat to supplies and reducing Russia's economic and strategic influence in the region. Since 1991, Washington has perceived Moscow both as a competitor and as a partner in the Caspian region.

Fourth, the U.S. sanctions policy against Iran expands to the Caspian. Washington has sought vigorously to isolate Iran and undermine Tehran's efforts to play a role in developing the region's energy infrastructure. Washington's rationale is that aside from sanctions, there could be security of supply concerns related to oil transported via Iran. Caspian oil sent via the Persian Gulf would be subject to the same potential bottleneck as much of Middle Eastern crude if flows were disrupted through the Strait of Hormuz. Thus, in spite of significant political changes in Tehran since 1997 (the election of President Muhammad Khatami), there has been no relaxation of U.S. anti-Iran Caspian policy. Strong opposition to an Iranian role in the Caspian faces two important challenges. U.S. oil companies believe that Iran provides the easiest, fastest, and cheapest route to transport Caspian oil and gas. This indicates that there is a contradiction between U.S. commercial interests and strategic ones. Also, some analysts suggest that Iranian inclusion in U.S. pipeline proposals would underscore the differences between Tehran and Moscow and intensify the competition between them.

Finally, the U.S. policy in the Caspian aims at supporting Turkey, a NATO member, to play a leading role in developing the energy infrastructure in the Caspian region. Shortly after the collapse of the Soviet Union, Washington urged the newly independent states to adopt the "Turkish model" of secularism, Western democracy, and free market.

To sum up, since 1991 the United States has played an active role in the Caspian region.

(1) U.S. oil companies, along with other Western and Russian companies, have taken the lead in developing Caspian oil and gas resources. Indeed, Chevron is the leading oil company in the Caspian, and the new merged company ChevronTexaco is well positioned to dominate oil and gas reserves and production ranking in the Caspian for several years.
(2) Over the years, the relations between Washington and Moscow with regard to the Caspian have become less confrontational and more accommodative.
(3) Given Iran's location and historical and cultural ties with the Caspian states, Tehran has been able to maintain and develop political and economic cooperation with Baku, Astana, and Ashgabat. Still, it can be argued, American sanctions have slowed down and weakened the Iranian influence in the region.
(4) On the other hand, American support to Ankara has contributed to strategic and economic partnership with Caspian states, particularly Azerbaijan.

Russia

Russia has strategic and commercial interests in the Caspian region. Some in the Russian elite see American penetration of Central Asia (what some perceive as Russia's backyard or the near-abroad) as an effort to displace Russia and marginalize its influence. Accordingly, Moscow has sought to use military and economic leverages to reassert its domination in the region and to contain Western influence in the region. This is particularly apparent in Moscow's role in mediating an end to military confrontations between and within Central Asia and the Transcaucasus states. It is important to point out that Azerbaijan is one of few former Soviet republics that have rejected any Russian military presence on their territory.

Another significant Russian interest in the Caspian is the region's hydrocarbon resources. The oil and gas industry occupies a prominent position in the Russian economy because it generates a substantial share of foreign revenues. Thus, Russia's energy relations with the Caspian states have a direct impact on its national economy. Moscow can influence oil and gas industry in Baku, Astana, and Ashgabat in a number of ways: as an investor or partner in field development and pipeline projects; as a transit country for their exports to formal Soviet republics and other markets; as a competitor in most of these markets; and as a market in its own right.[25]

In other words, clear economic dependency remains between Russia and the Caspian states. Moscow has not hesitated to use all these methods to serve and optimize its strategic and commercial interests in the Caspian, particularly since Vladimir Putin became president in 2000. Putin has always articulated a more coherent and aggressive Caspian policy than his predecessor, Boris Yeltsin. He urged the industry not only to become more involved in the Caspian energy development but also to take a more cooperative stance toward investment from other countries to help promote a higher level of total development. Some analysts believe that Russia's Caspian policy under Putin moved away from trying to contain U.S. expansion in the region in favor of "constructive engagement" with American government and oil companies.[26] In line with this more assertive Russian stance in the Caspian, the country's biggest oil and gas companies—Lukoil, Yukos, and Gazprom—teamed up and formed the Caspian Oil Company in 2000 to "help Russia strengthen its stand in the region."[27] Furthermore, in January 2002, Putin proposed the creation of a "Eurasian Alliance of Gas Producers," which would bring together Russia and the three big gas-producing countries of Central Asia, Kazakhstan, Turkmenistan, and Uzbekistan. Such an alliance would help guarantee Russia close, long-term economic ties with these countries, at a time when many Russians are worried by growing U.S. influence in the region as a result of the war in Afghanistan.

Three conclusions can be drawn from Moscow's policy in the Caspian since the early 1990s. First, Russia has many more economic and strategic levers in the region at its disposal than the United States or any other power. These include using military force or obstructing pipelines. These options are cheaper and easier for Russia than they are for the United States or other external powers. In other words, Moscow can easily play the role of spoiler if developments in the region do not go its way. Second, Russia's Caspian policy is increasingly driven by pragmatic considerations rather than strategic ones. Two examples illustrate this trend: the growing energy cooperation between Russia and its traditional enemy, Turkey, and the rising partnership in several oil schemes between Moscow and its top rival, Washington. Finally, Russia's pragmatic policy seems to be paying off. Since the late 1990s Moscow has succeeded in negotiating and constructing a number of pipeline schemes between its ports and oil and gas fields in the Caspian. Furthermore, Russian oil and gas companies are heavily involved in most multilateral energy consortia in the region.

Turkey and Iran

Since 1991, Turkey's policy in the Caspian region has been driven by historical ties as well as economic interests. Traditionally, the region had a cultural Turkic and Persian heritage. However, fifty years of Russian rule, followed by seventy years of Soviet domination, decisively isolated the Central Asian states from both Turkey and Iran. Following the collapse of the Soviet Union, Ankara sought to renew and reactivate these historical ties. With very few friends and allies in either the Islamic world or the European Union, some members in the political elite in Ankara perceive Central Asia as Turkey's strategic depth. No wonder Turkey granted diplomatic recognition to most of these states shortly after the collapse of the Soviet Union and opened embassies almost everywhere in Central Asia. These diplomatic ties reflect strong Turkish interest in the region's energy resources. Turkey's domestic oil and gas production satisfies very little of its energy needs. Furthermore, the country's energy consumption is growing much faster than its production (particularly before the severe economic crisis of the early 2000s), making Turkey a rapidly growing energy importer. Finally, Turkey's strategic location makes it a natural "energy bridge" between major oil and gas suppliers from the Persian Gulf and Caspian Sea on one side and consumer markets in Europe on the other side. Thus, Turkey has sought to promote itself as the most appropriate route to transport Caspian oil and gas to the European markets. In addition, Turkish companies are actively engaged in the exploration and development of the region's hydrocarbons.

Similar to Ankara, Tehran saw the breakup of the Soviet Union as presenting both new threats and new opportunities. The risks that are commonly cited by Iranian analysts are that secessionism and extremist ethnonationalism or the weakness of state and nation-building processes and the weakness of political institutions could lead to state failure or civil war. This in turn could lead to interstate conflicts and a more general regional crisis, as well as generating a humanitarian crisis and refugee flows.[28] On the positive side, Iran's geographical location between Central Asia and the Persian Gulf and its relatively developed internal transport infrastructure offered the potential to play an important role as a bridge to the world economy for the land-locked Caspian states. There is no doubt that Tehran figures prominently in any discussion of current or future energy exports from the Caspian. It offers the simplest, shortest, quickest, and cheapest route to global markets for oil from the Caspian republics.

Given these risks and opportunities and Iran's experience in the Persian Gulf, Tehran has been particularly concerned about two possible threats —being excluded from a future Caspian Sea grouping (similar to the Gulf Cooperation Council in the Persian Gulf)[29] and the danger of hostile foreign penetration into the region (American military presence in the Persian Gulf). In order to offset these possible threats and take advantage of potential opportunities, Tehran embarked on an active and pragmatic strategy in the Caspian region. Like Turkey, Iran was among the first countries to grant diplomatic recognition to most of the new states in Central Asia following the breakup of the Soviet Union and to try to revitalize the traditional cultural and commercial ties with the region. The expansion of the Economic Cooperation Organization in 1992 to include Azerbaijan, Kazakhstan, and Turkmenistan (as well as Afghanistan, Kyrgyzstan, Tajikistan, and Uzbekistan) can be seen as an important step in this direction.

Several characteristics can be identified in Tehran's strategy in the Caspian region and Central Asia. First, since the early 1990s Iran has invested few resources in exporting Shi'ism or igniting militant Islam in the region. For example, Iran maintains better relations with Christian Armenia than with the predominantly Shi'ia Azerbaijan. Similarly, in Chechnya, Tehran has lent very little support, if any, to the Muslim rebels in their fighting against the Russian army since the mid-1990s. These two examples suggest that Islam is not the driving force for Iranian policy in the region. Instead, nationalist and commercial interests guide Tehran's strategy. Second, like Turkey and Russia, Iran has sought to play an active role in developing the Caspian's hydrocarbon resources through unilateral initiatives, bilateral agreements, and multilateral forums. In 1998, Tehran formed a consortium with Royal Dutch/Shell and Lasmo to develop oil and gas fields on its Caspian shores. Tehran has negotiated several swaps and pipelines schemes with Ashgabat, Astana, and Baku, and despite some political and border disagreements, Iranian entities have taken stakes in some international consortia to explore and develop oil and gas fields in these three Caspian states.

Third, Iran has repeatedly voiced its concern over an objection to the militarization of the Caspian Sea. Again, the similarities between the Caspian Sea and the Persian Gulf are striking. These include large oil revenues, authoritarian governments, and ethnic and religious tensions. These characteristics can lead to an arms race in the region. In 2001, Azerbaijan bought some patrol boats from the United States and Turkmenistan bought some from Ukraine and negotiated an arms deal with Russia. Iran is particularly concerned about the presence of the Russian navy in the Caspian. Given the military asymmetry between Russia and the other littoral states, Moscow will undoubtedly be the main beneficiary if an arms race is ignited. Fourth, in spite of this important disagreement between Moscow and Tehran, the two countries are united in their opposition to what they perceive as American penetration. Heavy investment in the Caspian states and a potential American military role (directly or through a third party such as Turkey) are seen as an attempt by Washington to undermine and reduce Russian and Iranian influence in the region. Accordingly, Moscow and Tehran have forged a strategic alliance to resist what they perceive as "American hegemony" in the Caspian and worldwide. Russia's huge military sales to Iran illustrate this growing military and strategic cooperation between the two countries.

Fifth, since the early 1990s, Ashgabat, Astana, and Baku have expressed interest in transporting some of their oil and gas resources through Iran. In addition to obvious advantages that Iran has (cheap and quick routes), these three countries are interested in diversifying their export pipelines. Thus, Turkmenistan, which was under gas monopoly by Russia, started exporting some of its gas to Iran in the late 1990s. In 2000, Kazakhstan commissioned France's TotalFinaElf, BG International of the United Kingdom, and Agip of Italy to carry out a feasibility study for a crude pipeline through Turkmenistan and Iran.[30] Azerbaijan, which is under heavier U.S. pressure and influence than Kazakhstan and Turkmenistan, has shown more interest in exporting its oil and gas through Turkey, Georgia, and Russia than the Iranian route.

In spite of this slow and reluctant cooperation between Iran on one side and Azerbaijan, Kazakhstan, and Turkmenistan on the other side, it is important to point out that in the long term, the two sides might be locked in an intense competition. The four economies are heavily dependent on oil and gas exports.

September 11 and Its Aftermath

The terrorist attacks against the United States on September 11, 2001, and the American war against terrorism that destroyed the Taliban regime in Kabul have introduced new political, military, and strategic parameters in Central Asia. This war is likely to have a significant impact on the region's energy policy, particularly in two areas: increasing interest in developing the Central Asia/Caspian Sea hydrocarbon resources and the revival of the southern route (transporting oil and gas from the Caspian fields to Pakistan through Afghanistan).

A common reaction in the West to the September 11 terrorist attacks has been what might be called "a curse on all their houses."[31] In other words, the Middle East is being increasingly presented and perceived as an unreliable source of oil and gas to the global market. The fact that private money from Saudi Arabia and other Gulf monarchies was used to fund terrorist organizations has intensified the anti-Arab feeling in the West and underscored the need to reduce dependence on oil supplies from the Persian Gulf. Furthermore, given their domestic constituencies, Gulf leaders (particularly Saudis) have been reluctant to provide strong, unconditional public support to the war on terror. On the other side, most governments in Central Asia have been much more forthcoming in supporting American military operations in Afghanistan. Several senior U.S. officials, including Secretary of State Colin Powell and Secretary of Defense Donald Rumsfeld, visited Central Asian states to cement and coordinate military and strategic cooperation. In short, in the aftermath of September 11, Central Asia/Caspian Sea states have proven themselves a strategic and reliable partner to the United States.

The second important impact of the war in Afghanistan on energy policy in Central Asia is related to pipeline construction. Afghanistan itself is poor in hydrocarbon reserves. Its significance stems from its geographical location as a potential transit route for oil and gas exports from Central Asia to the Arabian Sea. Unocal Corporation, a U.S. oil giant, proposed the construction of an 890-mile, $2 billion gas pipeline connecting Turkmenistan and Pakistan via Afghanistan. An agreement between the governments in Islamabad and Ashgabat and the Taliban to arrange funding for this proposal was signed in January 1998. Besides the gas pipeline, Unocal also had considered building an oil pipeline to carry oil from Turkmenistan to Pakistan's Arabian Sea coast via Afghanistan. In August

1998, however, U.S. embassies in Kenya and Tanzania were bombed and Washington retaliated by launching a missile attack against Osama bin Laden's camps in Afghanistan. In response to these developments, Unocal concluded that its pipeline proposals were no longer tenable as long as the Taliban were in power. The destruction of the Taliban regime in Kabul has removed this obstacle. Oil and gas pipelines from the Caspian Sea to Pakistan, India, and the Arabian Sea via Afghanistan are likely to be reconsidered once a sense of stability is established in Kabul.

To sum up, since the collapse of the Soviet Union in 1991, the United States, Russia, Turkey, and Iran have sought to advance their strategic and commercial interests in the Caspian Sea. Three conclusions can be drawn from this power jockeying. First, the interests of these four contenders are neither mutually exclusive nor identical. Increasingly, more policymakers in the four capitals have come to realize that the competition in the Caspian Sea should not be put in zero-sum terms and that cooperation could accelerate and reduce costs for the exploration and development of the region's energy resources. In other words, there are signs that economic considerations are gaining ground at the expense of political and strategic ones. Second, there is one exception for the previous conclusion—U.S./Iranian rivalry in the Caspian. In spite of some relaxation in the tension between the two countries since the election of Khatami in 1997, Washington has been persistent in opposing any Iranian role in the region. America's uncompromising stand has denied Iran the opportunity to realize the full advantages of its historical ties and strategic location. In other words, without American obstruction, Iran's role in the exploration and development of the Caspian resources would have been much bigger and more effective than it is. The Iranian role in developing the region's hydrocarbon resources has been reduced by U.S. opposition. Finally, the intense rivalry between Washington, Moscow, Ankara, and Tehran has slowed down the full utilization of Caspian resources. Exploration and development operations have been further complicated by internal problems.

Ethnic Divisions

Central Asia suffers from deep-rooted ethnic divisions. The Soviet iron rule put a temporary halt to hostility between the ethnic groups in the region. The breakup of the Soviet Union in 1991, however, has opened the door for claims and counterclaims by several sectarian and ethnic movements to resurface. Since the early 1990s, hostility has resumed within and between several Central Asian and Transcaucasus ethnic groups. From the outset, two general characteristics of this fragile balance between various ethnic groups in the region can be identified. First, most of these disputes are based on two competing principles: the territorial integrity of any state and the right of nations to self-determination. Both are registered in the key United Nations documents and recognized by international law. Second, these disputes have threatened and complicated oil and gas exploration and development operations in the Caspian states as well as the construction of transportation routes.

One of the most ambitious schemes is the one designed to carry oil from Baku to Ceyhan, the Turkish port on the Mediterranean Sea. Plans for the pipeline may be complicated by Ankara's struggle with Kurdish separatists. For most of the 1990s, oil company officials had confirmed that the possibility of a Kurdish attack on and sabotage of oil installations and pipelines was one major factor, among others, that they would consider in making decisions on transportation routes. However, Kurdish military opposition has abated since Ankara succeeded in arresting the Kurdish leader Abdullah Oclean in 1999. From his cell in a Turkish high security prison, Oclean announced a "peace initiative," ordering members of the Kurdistan Workers' Party to refrain from violence and requesting dialogue with Ankara on Kurdish issues.

Another factor that has adversely affected oil development and export is the war in Chechnya, which has obstructed Russia's efforts to promote a pipeline through the region. The original northern route for oil from Azerbaijan passed through the Russian republic of Chechnya en route to the Black Sea port of Novorosiisk. Russian troops entered Chechnya in December 1994, and after almost two years of fighting, a peace agreement was reached. Deadlocks over negotiations, however, prompted Moscow to build a bypass pipeline around Chechnya, heading north out of Azerbaijan via the southern Russian republic of Dagestan. The pipeline was completed in 2000, but various disputes have limited flow through the line. In addition, Dagestan has security concerns of its own. In August 1999, fighting flared between Islamic militants (largely from Chechnya) and Russian forces in Dagestan. In response, the Russian army invaded Chechnya in late September 1999. These Russian military operations have ensured years of bitter hatred from Chechens and underscored the continuing threats to oil pipelines through the region.

The western route for oil from Azerbaijan goes from Baku to the Georgian port of Supsa on the Black Sea, and several other proposed pipeline routes also pass through Georgia. These current and proposed pipeline routes pass near several regions of Georgia that have been the site of separatist struggles, such as Abkhazia (northwest Georgia) and Ossetia (north central Georgia). Shortly after the collapse of the Soviet Union, Tbilisi sought to expand its control over Abkhazia. This triggered a civil war, and in September 1993, Abkhazians, supported by Russia, defeated the Georgian forces and restored control over their entire territory. The protracted talks under the UN aegis with Russia acting as an intermediary produced a Cease-fire and Separation of the Forces Agreement, signed in Moscow in May 1994.[32] Talks have continued to resolve the standoff, including proposals to route future oil pipelines across Abkhazia, on the premise that economic cooperation could help bring peace to the region. Since the late 1990s, proposals to arrange for a special peacekeeping force to protect the oil export pipelines have been under consideration. These proposals have been negotiated with other members in the Partnership for Peace Program and GUUAM Group. Developments in South Ossetia are similar to those in Abkhazia. Again, with Russian help, the Ossets defeated Georgian troops and pushed them out of their territory. In 1992 peacekeeping forces, made up of Russian and Georgian troops, were deployed in the region. The outcome of these two ethnic confrontations in Georgia can be described as a state of "no peace, no war." This uncertainty has an adverse impact on oil and gas shipments through the region.

Another potential threat to the delicate ethnic makeup in the Caspian region is between Tehran and Baku. The Azerbaijani population has been divided between the independent state of Azerbaijan, with an estimated population of 7 million people in 2000, and more than 20 million Azeri in the northwestern provinces of Iran, approximately one third of the Iranian population. The Azeri question came to the fore in 1945–46 when the Soviet Union established an autonomous government of Azerbaijan in northern Iran in an attempt to expand its control. This failed attempt was short-lived, and the Azeri citizens of Iran continued to be assimilated in the Iranian economic and political systems. To a certain extent, the Azerbaijani elite has always been important in Iran from the Safavi and Qajar dynasties, to the Pahlavi regime and under the Islamic Republic. Unlike the Kurdish Iranians, there has not been a separatist movement among the Azeri Iranians. Thus, when expressions of Azerbaijani nationalism began to emerge in the late 1980s, in support of the movement emerging in northern Azerbaijan, this took the form not of a statement of separatism or of autonomy from Tehran but of a demand for greater support by Tehran for Baku vis-à-vis the Soviet Union and Armenia.[33]

In spite of this assimilation of the Azeris within the Iranian socioeconomic and political fabric, there is a growing concern over the so-called Turkic nationalism. Tehran's concerns were particularly acute during the tenure of nationalist Azerbaijani president Abulfaz Elchibey, who took over from former Communist Party leader Ayaz Mutalibov in 1992. Since Heydar Aliyev took office in 1993, Azerbaijani nationalist sentiments have been restrained. Still, there are some speculations on a potential impact of a rich independent Azerbaijan on the makeup of Iran's ethnic fabric and the overall relations between the two countries.

Azerbaijan is also the theater for the bloodiest ethnic conflict in the Transcaucasus, which involves neighboring Armenia. The conflict between the two former Soviet republics revolves around control of Nagorno-Karabakh, a remote mountainous enclave inside Azerbaijan that is populated principally by ethnic Armenians who want to become either independent or a part of Armenia. The western route for oil from Azerbaijan passes just north of Nagorno-Karabakh.

The conflict between Azerbaijan and Armenia is deeply rooted in the past. It became especially acute in late 1980s with the apparent weakness of the Soviet Union. In late 1989, in a joint session with the Nagorno-Karabakh National Council, the Armenian Supreme Soviet declared Armenia and Karabakh to be a "United Armenian Republic."[34] Following this declaration, several steps were taken to consolidate the emergence of the enclave as an independent state in the early 1990s. This led to a war between the Armenian and Azerbaijani communities there that was supported by both Baku and Yerevan. As a result, more than 30,000 people were killed and approximately 1 million Azerbaijani civilians were driven from their homes and became refugees. In addition, Armenian troops seized Nagorno-Karabakh and hundreds of square miles of Azerbaijani territory (about 20 percent).

This bloody war prompted the international community to try to put an end to the hostility and bring the two sides to the negotiations table. In 1992 the Organization for Security and Cooperation in Europe created the Minsk Group, chaired by Russia, France, and the United States, to mediate between the two parties. In 1994, Russia, backed by the Minsk Group and the United Nations, negotiated a cease-fire. This cease-fire stopped the fighting but has not offered a permanent solution to the conflict. Aliyev has offered to build a pipeline through Armenia en route to Turkey, which would give Yerevan transit revenues from the pipeline, in exchange for Armenian withdrawal from the occupied territories.[35]

The large gap between what the Armenians want and what their Azerbaijani counterparts want explains how the two sides have failed to find a common ground since 1994. The Armenians in Nagorno-Karabakh want to join Yerevan either de jure or, at least, de facto and would like their representatives to be invited to take an equal part in the negotiations. On the other hand, the Azerbaijanis want to restore their state sovereignty over the territories controlled by the Karabakh Armenians and their refugees to return home. In exchange, Baku promises the region a wide autonomy within Azerbaijan. The situation was further complicated in the early 1990s by the lack of support Baku received from either Moscow or Washington. In addition to the historical ties between Moscow and Yerevan, Russia negotiated and signed agreements with Armenia, under which the former became the main supplier of arms and fuel to the latter. On the other hand, the United States passed section 907, which prohibits U.S. assistance (with the exception of assistance for nonproliferation and disarmament programs) to the government of Azerbaijan under the Freedom for Russia and Emerging Eurasian Democracies and Open Markets Support Act of 1992 (also known as the Freedom Support Act) until the president determines that the government of Azerbaijan is taking demonstrable steps to cease all blockades and other offensive uses of force against Armenia and Nagorno-Karabakh.[36] This step was taken under heavy pressure by the powerful Armenian lobby. In 1998, modifications in the legislation permitting exemptions of humanitarian and nongovernmental organizations' assistance were approved.

Since the late 1990s, substantial efforts have been made to achieve a permanent peace between Azerbaijan and Armenia. These renewed efforts can be explained by at least four developments: First, Azerbaijan and Armenia are sandwiched between Russia, Turkey, and Iran. Any renewal of hostilities could draw in these powers as well as the United States. Second, since the early 1990s, several oil and gas discoveries have been made in the Caspian Sea, and pipelines plans have been drawn. Some routes to transport these deposits pass across or near the Azeri-Armenian borders. There is a growing need to secure the safety of these schemes. Third, President Heydar Aliyev, in his late seventies, is eager to settle the conflict to boost the succession chances for his son, Ilham. Finally, in spite of its wealthy diaspora, Armenia has not been able to attract foreign investment without the guarantee of a peace settlement.

Consequently, the chances for peace have improved markedly. The mediation effort has a highly unusual level of personal involvement by the presidents of the United States, France, and Russia, each of whom met with his Armenian and Azerbaijani counterparts in late 2000 and early 2001 to urge a spirit of compromise. President Robert Kocharian of Armenia and Heydar Aliyev of Azerbaijan held several face-to-face negotiations. Still, the two leaders are under heavy pressure from domestic opposition parties not to compromise.

To sum up, the Caspian region is overburdened by ethnic rivalries and struggles. In the early 2000s, most of these ethnic disputes are frozen but not resolved. Several factors have contributed to this outcome. These include the legacy of the Soviet mismanagement of the nationality issue, cynical manipulation by outside forces,[37] the sectarian and ethnic animosity between different groups within the region, as well as the slow process of nation-building. An end to ethnic hostility within and between the Caspian states and the achievement of a lasting peace need an endorsement by all regional powers. The same conclusion can be applied to a final agreement on the legal status of the Caspian Sea.

The Legal Status of the Caspian Sea

Uncertainty surrounding the legal regime that will eventually govern the oil and gas exploration and development operations in the Caspian Sea is a major risk that investors have had to take into consideration in doing business in the region. Until the breakup of the Soviet Union in 1991, this problem (lack of agreement by the five littoral states on how to divide the sea between them) simply did not exist, or more accurately, it was at a much lower level. The main quarreling between Russian and Persian ships in the nineteenth century was over fishing and trade routes. Mineral deposits, at the bottom of the sea, were not known yet. Later, the two states signed two documents to govern their relationship in the Caspian Sea— the Friendship Treaty of 1921 and the Treaty of Commerce and Navigation of 1940. Some of the main points of these two treaties need to be highlighted.

First, Moscow and Tehran agreed that the Caspian was only open to their own vessels. This confirmed the exclusivity and equality of rights of the two coastal states to the Caspian and that the sea was effectively closed to the rest of the world.[38] Second, the two treaties did not differentiate between warships and other types of ships (passenger and transport). Third, the two sides reserved a twelve-mile zone along their respective coasts for the exclusive fishing rights. Still, no attempt was made to delimit any official sea boundary between them. Fourth, the two treaties did not deal with some important issues such as protecting the natural environment of the sea and the development of mineral deposits at the bottom of the sea. In short, the 1921 and 1940 treaties between Moscow and Tehran suffered from some ambiguity and left important questions unanswered. Accordingly, many analysts and policymakers have questioned the applicability of these two documents to the new situation in the Caspian since the early 1990s.

Following the signing of an investment agreement between the Azeri government and a BP-led consortium (later known as the Azerbaijan International Operating Company) in 1994, the need to determine the legal status of the Caspian has become a crucial issue for all concerned parties (the five littoral states and foreign companies). Moscow warned that the issue of ownership of Caspian resources needed to be settled before any individual state starts exploration and development operations. The implication of Russia's claim was that it or any other littoral state might be able to block or critically delay projects, either by veto or by requiring a cumbersome approval process.[39] The Azeri government responded by arguing that the two bilateral treaties between Iran and the Soviet Union did not deal with mineral resources in the Caspian and therefore should have no bearing on such a discussion. Moreover, during the Soviet era Moscow treated the Caspian as if it were divided into two sectors and did not consult Tehran on development outside its twelve-mile zone. Finally, Baku claims that the Soviet sector of the sea was divided into economic zones, related to the land boundaries of the republics. In short, Azerbaijan believes that it has historical and legal rights to divide the Caspian Sea into national sectors and proclaim its sovereignty over its share.

The United States perceived this step by Russia as an attempt to make Caspian projects less attractive to foreign investors and, in turn, hamper the economic development of Azerbaijan, Kazakhstan, and Turkmenistan, which hoped to attract significant investment to their offshore regions in order to consolidate their economic and political independence from Moscow. The Russian initiative, however, has not succeeded in blocking joint adventures between international oil companies and the newly independent Caspian states despite deep disagreement on the legal status of the sea.

A fundamental question in this debate on the legal status of the Caspian is whether it is a sea or a lake. According to the UN Convention on the Law of the Sea, nations bordering a sea may claim twelve miles from shore as their territorial waters and a 200-mile Exclusive Economic Zone beyond that.[40] If the Law of the Sea were applied to the Caspian, full maritime boundaries of the five littoral states bordering the Caspian would be established based upon an equidistant division of the sea and undersea resources into national sectors. If the Law of the Sea were not applied, the

Caspian and its resources would be developed jointly—a division referred to as the condominium approach. The five littoral states have not agreed on whether the Caspian is a sea or a lake. This lack of consensus is due to the conflicting national interests between all the concerned parties.

In spite of original opposition to dividing the Caspian into national sectors, Russia's position significantly changed in late 1996. At a meeting of foreign ministers of the five Caspian states, Moscow proposed that within a forty-five-mile coastal zone each country could exercise exclusive and sovereign rights of the seabed mineral resources. The central part was to remain common property with its hydrocarbon resources developed by a joint-stock company of the five states. Both Azerbaijan and Kazakhstan rejected the Russian proposal. Meanwhile, Moscow, Tehran, and Ashgabat tried unsuccessfully to set up a tripartite company to develop seabed hydrocarbon resources. Turkmenistan preferred to open an international tender on the resources along its coast where the tripartite company was supposed to work.[41]

In 1998, there was another important turn in the Russian position. After long negotiations with Kazakhstan, the two countries signed an agreement dividing the northern Caspian seabed alone along median lines between the two countries, with the waters (covering issues such as shipping, fishing, and environment) remaining under joint ownership.[42] In January 2001, Russia signed a similar agreement with Azerbaijan. In spite of Russia's endorsement of dividing the Caspian's seabed into national sectors, uncertainty still characterizes Moscow's stand. After a summit between Putin and his Iranian counterpart, Khatami, in March 2001, the two leaders issued a statement confirming that until the legal regime of the Caspian Sea is finalized, the two countries do not officially acknowledge any boundaries there. Furthermore, agreements on the legal status of the sea will only be valid if they are approved by all five littoral states.[43] In other words, the position Moscow adopted with Tehran suggests some contradictions with the two agreements it signed with Astana and Baku. Another Russian proposal is to split the Caspian into northern and southern sections. The former would be divided between Moscow and Astana, while the latter would be divided equally between Tehran, Ashgabat, and Baku. Azerbaijan opposes this proposal because it would boost Iran's share at its expense.

The evolution of Moscow's stand on the legal status of the Caspian from endorsing joint management to gradually accepting national sectors suggests three conclusions. First, generally speaking, the Russian Ministry of Foreign Affairs has opposed the division of the Caspian into national zones while Russian oil companies have participated in lucrative schemes to develop oil and gas fields in Azerbaijan, Kazakhstan, and Turkmenistan. Besides, Russia has sought to develop hydrocarbon resources within its own sector of the sea. In other words, Russia's economic interests seem to have gained the upper hand in formulating the country's Caspian policy at the expense of strategic interests. Second, Russia has opposed the laying of trans-Caspian pipelines until a legal framework is established to govern environmental and biological issues and to establish legal responsibility for safe use of the Caspian Sea.[44] Third, Russia has asserted that the airspace above the Caspian, the surface of the sea, and the waters of the sea should remain open and be administered jointly while the floor of the Caspian can be divided roughly along median lines among the littoral states. Given the asymmetry in military and naval power between Russia and the other four Caspian states, this approach (dividing only the seabed) gives Russia significant strategic advantages.

Not surprisingly, Iran, the second most powerful country in the Caspian, strongly opposes the Russian approach. Indeed, the Caspian is the main issue bedeviling Russian-Iranian relations, despite a number of breakthroughs on sales of arms and nuclear technology. Three characteristics can be identified in Iran's stand on the legal question of the Caspian:

(1) After the collapse of the Soviet Union, Iran maintained that the two treaties signed by the two countries in 1921 and 1940 should be considered valid and should govern the Caspian until the five littoral states sign a new one. In other words, originally, Iran favored a "condominium approach" meaning joint management of the Caspian's resources.
(2) Faced with strong opposition by the other Caspian states to joint ownership, Tehran reluctantly and gradually has expressed willingness to accept a system of division that would leave it with a share of not less than 20 percent. The Iranian position is to "divide all equally or divide nothing."
(3) If the legal regime is to divide, then both the seabed and the surface should be divided equally.

Iran insists on an equal share with the other Caspian states because a simple division of the sea into national sectors based on the median line would give it the smallest share.[45] Should this method be modified to include other factors such as the length of the shore and the convexity and the slope of the seabed near the shore, Iran's share of the Caspian Sea would be 13.6 percent, Turkmenistan 18 percent, Russia 19 percent, Azerbaijan 21 percent, and Kazakhstan 28.4 percent.[46] Furthermore, Iran's sector of the sea is not thought to be rich in hydrocarbon resources.

In spite of Tehran's opposition to dividing the Caspian into national sectors, Iranian oil companies, like the Russian ones, have participated in projects to develop oil and gas fields in the other Caspian states. In addition, Iran is working with foreign oil companies to develop its own oil and gas deposits in the Caspian. Iran's claimed fields, however, overlap with those claimed by Azerbaijan.

In order to understand the Azeri, Kazak, and Turkmen stand on the legal status of the Caspian, two facts need to be highlighted. First, if the Caspian were divided into national zones, the largest hydrocarbon deposits would be in the Azeri and Kazak sectors and to a lesser extent in the Turkmen sector. Russia and Iran are believed to have fewer resources in their sectors than those of the other three Caspian states. Second, in comparison with Azerbaijan, Kazakhstan and Turkmenistan are more dependent on Russia. One third of Kazakhstan's population is Russian, and Turkmenistan relies extensively on Russia in exporting its gas. Thus Azerbaijan can afford and has adopted a more independent stand from Russia than those taken by Kazakhstan and Turkmenistan.

Baku has objected to Tehran's decision to award Royal Dutch/Shell and Lasmo a license to conduct seismic surveys in a region that Azerbaijan considers to fall in Azeri territory. Similarly, Baku and Ashgabat have been involved in a dispute over three fields called Kyapaz, Azeri, and Chirag in Azerbaijan, and Serdar, Khazar, and Osman in Turkmenistan. These disputes concern where to draw the median lines. Baku was the first and is the strongest proponent of dividing the Caspian (seabed, water, and surface) into national sectors.

Like Azerbaijan, Kazakhstan advocates a division of the Caspian. Accordingly, in 1997 Astana signed a communiqué with Ashgabat pledging to divide their sections of the Caspian along median lines. A year later, Kazakhstan signed a bilateral agreement with Russia dividing the seabed between the two countries. Meanwhile, Astana opposes Tehran's proposal to divide the Caspian equally among the five littoral states. Also, it is important to point out that despite Kazakhstan's strong support for the establishment of national sectors, it has stated that cooperation on the environment, fishing, and navigation would be beneficial to all littoral states. Finally, as Western investment has grown, Azerbaijan and Kazakhstan have become more assertive in affirming their territorial claims. During the CIS summit in Moscow (held in November 2001), the two countries signed a bilateral accord to divide the seabed along a modified median line. This agreement should be seen as an inevitable formality, since Kazakhstan agreed with Russia on the median line concept in 1998 and Azerbaijan followed suit in January 2001.

Turkmenistan's position has generally been somewhere between those adopted by Russia and Iran. Ashgabat has been less assertive than Astana and Baku in calling for a division of the Caspian into national sectors. Still, Turkmenistan's view has gradually shifted in favor of division. Turkmen leaders have repeatedly expressed the need for a unanimous agreement on the legal status of the Caspian by all five states.

In July 2001, an important development between Iran and Azerbaijan underscored the need for an agreement on the legal status of the Caspian. On July 23, an Iranian warship threatened to fire on an oil exploration vessel operated by the oil company BP/Amoco. The ship was conducting research near Alov deposit (Alborz in Persian), an offshore oil field that both Azerbaijan and Iran claim. After the Iranian intervention, BP officials told the Azerbaijan state oil company that it would not resume exploration of the area until the dispute was settled. Clearly BP was motivated by safety concerns and a desire to avoid antagonizing Iran. The United States called the Iranian action provocative, and Russia condemned the use of military threat.

This episode and the evolution of the five littoral states' stand on the legal status of the Caspian suggest several conclusions. First, the debate has shifted from choosing between dividing the sea into national sectors or joint ownership to where to draw the boundary lines. All five states agree on the principle of dividing the seabed. The disagreement is over how large the share of each state should be. Second, the legal uncertainties do not appear to have significantly slowed investment in the Caspian Sea. Still reaching an agreement would remove a significant stumbling block. Third, Caspian Sea demarcation should be viewed more as a political issue and less as a legal matter that still demands resolution. The contesting sides attempt to present demarcation of borders in the Caspian Sea as a strictly legal issue. However, the recurrent changes in their positions, as the above analysis shows, reflect that the actors' legal stances are tactical and that their overriding concerns regarding delineation are political and economic. Fourth, the Russian and Iranian participation in oil and gas exploration and development in the other three Caspian states suggests that a pattern of de facto benefit sharing seems to be developing. This is likely to reduce tension over the legal status of the Caspian. It might also contribute to alleviating disputes over pipelines.

Pipeline Diplomacy

The leaders of Azerbaijan, Kazakhstan, and Turkmenistan view the development of their hydrocarbon resources as a cornerstone to their economic prosperity. These states, however, are landlocked. They cannot ship their oil and natural gas by tankers from domestic ports. Instead, they have to be transported to the target markets by pipelines, which cross multiple international boundaries. Thus, the issue of potential routes through neighboring countries has become a priority for both regional and international powers, as well as for oil companies. The construction of a pipeline would provide the transit states with several financial and political benefits. These include access to oil or natural gas for their domestic needs; foreign investment and jobs; substantial transit fees; and political leverage over the flow of oil and gas.

The process of choosing and constructing pipeline routes is complicated and requires delicate negotiations with many parties. Until recently, the existing pipelines in the Caspian region were designed to link the former Soviet Union internally and were routed through Russia. Most of the Caspian's oil and gas shipments terminated in the Russian Black Sea port of Novorosiisk. This existing network, however, does not correspond to the new economic and political dynamics since the early 1990s. First, given the huge investment and efforts to develop the Caspian oil and gas potentials, there is doubt that the Russian network can meet the projected increasing level of exports. Second, in order to gain access to the European market, oil tankers from Novorosiisk have to traverse the crowded Bosphorus Strait, creating environmental and safety hazards. Third, this Russian network is aimed at the Mediterranean market; it does not target the vast Asian states. And fourth, there are political and security concerns as to whether these Caspian states should remain so dependent on Russia as their sole export outlet. Given these restraints, there is a growing consensus that multiple routes should be constructed.

For several years, a number of proposed routes have been under consideration. These include a pipeline to the north across Dagestan to Novorosiisk, completed in 2000, a second pipeline to the east from Kazakhstan and Turkmenistan to China; a third to the southwest through Afghanistan to Pakistan and India; a fourth to the south across Iran; and a fifth to the west, from Baku in Azerbaijan to the Georgian port of Supsa on the Black Sea, which became operational in April 1999, or the Turkish port of Ceyhan on the Mediterranea.[47] As has been discussed earlier, some of these routes pass through or near politically troubled areas, including Nagorno-Karabakh (Armenia-Azerbaijan), Abkhazia and Ossetia (Georgia), Chechnya and Dagestan (Russia), and Afghanistan. Since the 1990s, fighting in most of these conflicts has been suspended by cease-fires, not a genuine peace. Violence, terrorism, and instability can reemerge.

Given this uncertainty, the notion of multiple routes can enhance the security of oil and gas shipments from the Caspian region to the international market. Furthermore, several pipelines would promote economic competition and eventually reduce energy prices for consumers. However, these security and economic advantages are not without limitations. They have to be balanced by economic feasibility, since a larger number of pipelines would mean smaller economies of scale and greater expense for each project. In short, the process of selecting an export route from the Caspian Sea is complicated and involves several political and economic considerations.

For several years, international companies and the concerned governments have been engaged in serious negotiations to determine the priority of each pipeline. Since the late 1990s, attention has focused on four schemes. Two are supported by the United States: Baku-Ceyhan, and Trans-Caspian Pipeline (TCP), and two involve Russia: Caspian Pipeline Consortium (CPC) and Blue Stream. In the following, these four projects will be discussed as well as Iranian efforts to swap oil and gas with other Caspian states.

At least three strategic goals have shaped Washington's decision on choosing the most appropriate route for transporting the Caspian's oil and gas resources to international markets: first, to eliminate any Iranian role in Central Asia and the Caucasus; second, to strengthen economic and political ties between Turkey, a NATO member, and the three Caspian states (Azerbaijan, Kazakhstan, and Turkmenistan); and third, to support and promote economic prosperity and development in these newly independent states and help them to shape their own destinies.

Guided by these strategic goals, the United States has strongly supported two transportation options: One is to carry oil from Baku, Azerbaijan, on the western shore of the Caspian Sea, to Turkey's eastern Mediterranean oil terminal at Ceyhan; the other, the Trans-Caspian Pipeline, will transport natural gas from Turkmenistan to Turkey. The first scheme has been promoted by Washington as the main export pipeline; this will be some 1,080 miles long, with an estimated cost of between $2.4 billion and $3.7 billion.[48] The project is expected to transport about 1 million barrels a day and to be completed by 2004. A series of initial agreements that outlined the legal framework for the construction of this pipeline were signed in Istanbul in November 1999 by Turkish president Suleiman Demirel, Heydar Aliyev of Azerbaijan, and Georgian president Eduard Shevardnadze. President Clinton endorsed these agreements as a witness. In April 2000, the three countries signed the final document. Meanwhile, the Russian stand on this project has evolved from outright opposition, to competition, and finally to acquiescence.

Baku-Ceyhan Pipeline, however, faces a substantial obstacle: concern over the availability of sufficient volumes of crude. Not enough oil has been discovered offshore Azerbaijan to justify the construction of this costly pipeline. It needs 6 billion barrels of proven oil reserves to be commercially viable. The AIOC, created in 1994 to develop the Azeri-Chirag and Gunashli field complex in the Azeri sector of the Caspian Sea, has said that the field's reserves are estimated at 4 billion barrels.[49] Thus, the pipeline will require commitments from other producers and the United States sees Kazakhstan as a major potential contributor to Baku-Ceyhan. Recent reports from the Offshore Kazakhstan International Operating Company that a vast Kashagan oil field was discovered in the north Caspian have raised hopes that Kazakhstan could indeed commit significant volumes of crude to this route. Astana, however, has refused to make such a commitment, preferring to explore different options for exporting its oil. In December 2000, Kazakhstan commissioned TotalFinaElf of France, BG International of the UK, and Agip of Italy to carry out a feasibility study for a pipeline through Turkmenistan and Iran. Similarly, in March 2001, the construction of a pipeline from Kazakhstan to the Russian Black Sea port of Novorossiisk was completed.

The other leg of the U.S. pipeline strategy is the TCP, designed to tap Turkmenistan's world-class natural gas reserves. In February 1999, Ashgabat awarded a contract to build an export route to a consortium led by the U.S. firms Bechtel and General Electric Capital. This ambitious new link is planned to carry up to 30 billion cubic meters, starting in eastern Turkmenistan, crossing the Caspian seabed, and then passing through Azerbaijan and Georgia on its way to Turkey. Ankara is being viewed not just as a potentially booming gas market in its own right but also as the gateway to other consumers in southern and central Europe.

Things appeared to be going well for this project until July 1999, when BP/Amoco announced a major gas find at Shah Deniz, Azerbaijan, the first major offshore gas discovery in the Caspian Sea since 1991. As has been mentioned earlier, the field is estimated to contain between 25 tcf to 39 tcf of natural gas. Thus, the hurry to bring Shah Deniz gas on line is driven not so much by the obvious magnitude of the reserves as by the fact that, geographically, Azerbaijan is better placed than most other producers targeting the growing Turkish gas market, including Turkmenistan. Baku did not waste any time. In March 2001, it signed an agreement to supply Turkey with 3.1 tcf of natural gas over a fifteen-year period, starting in 2004.[50] A $1 billion pipeline will be built from Shah Deniz to the Turkish city of Erzrurm,[51] where it will connect with the Turkish gas transmission.[52] Another significant setback to the TCP occurred in November 2000. In the late 1990s, Turkmenistan was locked in a price dispute with Russia's giant Gazprom over the export of Turkmen gas. In 2000, the two sides were able to overcome their differences and reached an agreement under which Russia imported an estimated 1 trillion cubic feet of Turkmen natural gas. This sale accounted for most of Turkmenistan's gas production in 2000 and is seen by many analysts as a blow to the TCP. Finally, it is important to point out that Turkmenistan faces competition over the Turkish market not only from Azerbaijan's Shah Deniz but also from Russia's Blue Stream.

In December 1997, Moscow and Ankara signed an agreement to build a pipeline connecting northern Turkey with Russian gas fields via the Black Sea, with an estimated cost of between $2 billion and $3 billion. In order to secure the necessary funding, Gazprom formed a partnership with Eni S.p.A., the big Italian energy company, and its Saipem construction unit.[53] In 2000, work on the project started on both the Russian and Turkish sides. The capacity will start at 8 billion cubic meters a year (cm/y), building up to 16 billion cm/y in 2008.[54] Four geopolitical characteristics of this project can be identified. First, since the late 1990s Blue Stream and the TCP have reflected a renewed strategic and economic rivalry between Moscow and Washington. The former was initiated and is led by Gazprom, while the latter was strongly supported by the United States. Second, Blue Stream received strong support from big Turkish corporations doing business in Moscow. They saw it as a way to strengthen economic ties between the two countries. The Turkish government was less enthusiastic about it, given the country's strategic alliance with the United States. Third, approximately two thirds of Turkey's gas comes from Russia. Blue Stream will further deepen Ankara's energy dependence on Moscow. Finally, since the late 1990s and early 2000s, Turkey has agreed to import gas from several sources, including Turkmenistan, Azerbaijan,

Iran, and Russia. It is unclear whether Turkey can absorb all these gas deals, particularly in light of the economic crisis that has hit Ankara in the early 2000s.

In addition to Blue Stream, Russia has succeeded in advancing its interests on another frontier—the CPC. Unlike the Blue Stream and TCP, which together reflect rivalry between Moscow and Washington, the CPC is seen by many analysts as a good example of cooperation between the two countries. The 940-mile line connects the Tengiz oil field in Kazakhstan with the Russian Black Sea port Novorosiisk, with an initial capacity of 560,000 b/d, rising to 1.35 million b/d. Tengiz is one of the world's largest oil fields with substantial high-quality proven reserves. The American oil giant Chevron began negotiating a deal to develop the field in 1990 before the demise of the Soviet Union. TengizChevroil, a joint venture between Chevron, ExxonMobil, and Kazakhstan, became operational in 1993.[55] Originally, Russia was reluctant to ease its grip on a rival producer. Its attitude, however, changed as it found new ways for its interests to be defined. Russia became the biggest shareholder in the CPC with a 24 percent interest.[56] The first oil from the pipeline was scheduled to be loaded in June 2001, but several customs problems and technical hitches caused delays. After Russia and Kazakhstan reached agreement on transit tariffs, the pipeline was officially opened on November 27, 2001.

Two important projections can be made regarding the construction of the CPC. First, for the foreseeable future Russia will continue to be the main outlet for oil shipments from Kazakhstan. Second, given the limited capacity of the CPC and the anticipated expansion of Astana's oil production, it is unlikely that CPC will be enough to handle Kazakhstan's export surplus for long. Indeed, since the mid-1990s, the Kazak oil authorities have considered other options, including swap arrangements with Iran.

Given Iran's strategic location between the Caspian Sea and the Persian Gulf and its relatively advanced energy infrastructure, Tehran sees itself as a natural transit route for oil and gas exports from the landlocked Central Asian countries to world markets. Iranian officials strongly believe the southern route through Iran offers a much lower cost than existing routes to the Black Sea, especially those proposed to the Mediterranean. Iranian efforts to promote this vision have been restrained by strong U.S. opposition. As a result, since the mid-1990s much of the activities involving an Iranian role in exporting Caspian hydrocarbon resources have focused on swap arrangements. Simply, swaps mean that companies would ship oil from Azerbaijan, Kazakhstan, and Turkmenistan to the nearby Iranian ports on the Caspian and refineries in the northern part of the country. In return, the companies will pick up equivalent amounts of oil from Iran's terminals on the Persian Gulf. These arrangements appear to suit all involved parties. Most of Iran's oil fields are in the south, while population concentration centers are in the north. Thus, Tehran has to ship oil from the south to the north. Tehran would like to deliver Caspian crude to its refineries in the north in order to save on transportation costs. On the other hand, swap arrangements help oil companies sidestep U.S. sanctions because they do not involve any investment in Iran's energy sector.

Accordingly, under an arrangement made in the late 1990s, about 50,000 b/d are transported from Turkmenistan to the Tehran refinery through an existing pipeline from the northeastern Caspian port of Neka. An equivalent volume of crude oil, with quality differentials accounted for, is then swapped out from Iran's southern export terminal at Kharg Island. Tehran has an ambitious plan to expand its oil swap network with Ashgabat to 700,000 b/d.[57] Similarly, in 1996 Kazakhstan and Iran signed an agreement to begin oil swaps. The volumes involved were limited as a result of technical issues, including initial problems with processing Kazak crude in the Iranian refineries. Since then, Iran has sought to upgrade its refineries so they can treat the Kazak crude. The two countries plan to increase the volume of oil they swap. Finally, Iran has proposed to buy or swap oil from Azerbaijan. Baku has made its approval conditional on the settlement of disputes with Iran concerning the legal status of the Caspian and on progress toward improved relations between Iran and the United States.

Three important points need to be underscored regarding Iran's efforts to make swap arrangements and to construct pipelines from the Caspian oil fields to its export ports on the Persian Gulf. First, Iran faces difficulties securing the necessary financial resources to build the required routes to export Caspian oil. A centerpiece of Tehran's pipeline diplomacy is the construction of a 240-mile pipeline between the Caspian Sea port of Neka and Tehran. A local firm, Iran Power Plant Management Company, failed to raise the necessary $400 million. Consequently, the project was handed over to a Chinese consortium in February 2000. The European oil-trading company Vitol is involved in financing the project. Second, the United States has strongly opposed oil companies' use of swap arrangements with Iran. Indeed, two U.S. companies—Optimarket and Mobil—applied to participate in swap arrangements between Iran and other Caspian states, but their applications were turned down by the U.S. government. Washington objections, however, have not stopped European companies from pursuing swap arrangements.[58] Finally, despite American opposition,

Azerbaijan, Kazakhstan, and Turkmenistan, along with several international oil companies, consider the southern route through Iran as a vital and cost-effective option for exporting both their oil and their natural gas.

Three conclusions can be drawn from this discussion of pipeline diplomacy in the Caspian Sea. First, given the domestic, regional, and international rivalries surrounding oil and gas fields in the Caspian, there is no doubt that multiple export routes would increase the energy security for consumers, producers, and the global energy markets by making deliveries less vulnerable to technical or political disruptions on any individual route. Still, as pointed out earlier, energy security will have to be balanced by economic feasibility, since a larger number of pipelines would mean smaller economies of scale. Second, in many cases, particularly U.S. efforts to deny Iran a role in transporting Caspian oil and gas, the decision to choose the most appropriate route reflects a competition between strategic concerns and economic interests. Most pipelines are built by companies, not by governments. Ultimately petroleum companies are profit-driven entities accountable to their shareholders. They are in the business of making money, not in the business of achieving governments' strategic political objectives. In the long term, pipelines that make economic sense are more likely to be built than those that do not. Third, pipelines' capacity and availability will, to a large extent, influence the timing of oil and gas development in the Caspian region.

The Caspian Sea—What Lies Ahead?

Prediction is a risky adventure, particularly with regard to the Caspian region with its rich hydrocarbon deposits and ethnic divisions, as well as regional and international rivalry. Still, the foregoing analysis suggests several guidelines that are likely to influence energy policy in the region in the foreseeable future. First, the exploration and development of oil and gas deposits in the Caspian will take some time and tremendous efforts to realize their potentials. Complex technical, economic, logistical, geopolitical, and social obstacles will have to be overcome before the full utilization of these promising deposits. Second, the fluctuation of global oil prices will have a strong impact on exploration and development operations in the Caspian. Low oil prices would slow down these operations, given the high costs of production and transportation.

Third, since the influx of Caspian oil and gas into global markets will be slow and gradual, they are not likely to have a drastic impact on world energy markets and prices. The Caspian's growing production will contribute to the diversification of oil and gas supplies and consequently to global energy security. Still, the Persian Gulf will continue to occupy the driver's seat in meeting the growing world demand. Finally, the rivalry between the United States, Russia, Turkey, and Iran will not abate any time soon. Instead, the discovery of new oil and gas deposits will further intensify these strategic and commercial competitions. In order to ensure regional stability and accelerate full utilization of the region's hydrocarbon resources, all players have to be included. Excluding any state increases the chances that it might act as a spoiler. Inclusion, not exclusion, will ensure economic prosperity and energy security.

Glossary

The following glossary explains some of the technical terms that are used in this book or that readers are likely to encounter. It does not purport to be at all comprehensive.

Barrel of oil: Standard oil industry measure of volume. One barrel is equivalent to forty-two U.S. gallons.
Biomass: Organic nonfossil material of biological origin constituting a renewable energy source.
Brent blend: The principal grade of UK North Sea crude oil in international oil trading.
British thermal unit (Btu): The amount of heat required to increase the temperature of one pound of water by one degree Fahrenheit. One pound of coal has an energy content of between 10,200 and 14,600 Btu, while crude oil has an energy content of between 18,300 and 19,500 Btu/pound. One thousand cubic feet of natural gas has an energy content of 1 million Btu on average.
Carbon dioxide (CO_2): A colorless, odorless, nonpoisonous gas that is a normal part of the earth's atmosphere. Carbon dioxide is a product of fossil fuel combustion as well as other processes. It is considered a greenhouse gas, as it traps heat radiated by the earth into the atmosphere and thereby contributes to the potential for global warming.
Coal: A readily combustible black or brownish-black rock whose composition, including inherent moisture, consists of more than 50 percent by weight and more than 70 percent by volume of carbonaceous material. It is formed from plant remains that have been compacted, hardened, chemically altered, and metamorphosed by heat and pressure over geologic time.
Conventional oil: Crude oil that, at a particular time, can be technically and economically produced through a well, using normal production practice and without altering the natural viscous state of the oil. Nonconventional oil is more expensive to explore and develop, although there have been major cost reductions in the past few years.

Crude oil: Oil produced from an underground reservoir that has not been subjected to any refining or chemical process other than the separation at atmospheric pressure of any gases that were dissolved in the oil at the greater pressure of the reservoir.

Deregulation: The elimination of some or all regulations from a previously regulated industry or sector of an industry.

Downstream: That part of the petroleum industry that involves refinery, transportation, and marketing operations as contrasted with upstream operations of exploration, development, and production.

Dubai: A grade of crude oil which has effectively replaced Saudi Light as the "marker" crude oil in the Persian Gulf.

Energy source: A substance, such as oil, natural gas, or coal, that supplies heat or power. Electricity and renewable forms of energy, such as wood, waste, geothermal wind, and solar, are considered to be energy sources.

Fossil fuel: Any naturally occurring organic fuel formed in the earth's crust, such as oil, coal, and natural gas.

Gas to liquids (GTLs): A process that combines the carbon and hydrogen elements in natural gas molecules to make synthetic liquid petroleum products, such as diesel fuel.

Global warming: An increase in the near surface temperature of the earth.

Hydrocarbon: An organic chemical compound of hydrogen and carbon in the gaseous, liquid, or solid phase.

Liquefied natural gas (LNG): Natural gas (primarily methane) that has been liquefied by reducing its temperature to minus 260 degrees Fahrenheit at atmospheric pressure.

"Missing Barrels": Estimates of world oil inventories indicate that world oil supply/demand balances do not fully reflect the state of the world oil market. This implies that production has been overestimated, consumption data have been underestimated, or that the stock data are incorrect. This situation has become known as the "missing barrels" problem.

Natural gas: A mixture of hydrocarbon compounds and small quantities of various nonhydrocarbons existing in the gaseous phase or in solution with crude oil in natural underground reservoirs at reservoir conditions.

Natural gas liquids: A general term for all liquid products separated from natural gas in processing or cycling plants.

New York Mercantile Exchange (NYMEX): The most successful market for oil futures contracts on which very large volumes of heating oil and crude oil (WTI grade) in particular are traded. NYMEX has considerable influence on the physical trade.

Nuclear electric power: Electricity generated by an electric power plant whose turbines are driven by steam generated in a reactor by heat from the fissioning of nuclear fuel.

Nuclear reactor: An apparatus in which a nuclear fission chain reaction can be initiated, controlled, and sustained at a specific rate.

Organization of Petroleum Exporting Countries (OPEC): Was founded in Baghdad, Iraq, in September 1960, to unify and coordinate members' petroleum policies. Original OPEC members include Iran, Iraq, Kuwait, Saudi Arabia, and Venezuela. Between 1960 and 1975, the organization expanded to include Qatar (1961), Indonesia (1962), the United Arab Emirates (1967), Algeria (1969), and Nigeria (1971). Ecuador and Gabon were members of OPEC, but Ecuador withdrew in December 1992, and Gabon followed suit in January 1995. OPEC members produce about 40 percent of the world's oil, hold more than 77 percent of its proven oil reserves, and contain most of the global excess oil production capacity.

OPEC pricing: OPEC collects pricing data on a "basket" of seven crude oils—Algeria's Saharan Blend, Indonesia's Minas, Nigeria's Bonny Light, Saudi Arabia's Arab Light, Dubai's Fateh (or Dubai), Venezuela's Tia Juana Light, and Mexico's Isthmus (a non-OPEC crude oil)—to monitor world oil market conditions.

Pipeline: A continuous pipe conduit, complete with such equipment as valves, compressor stations, communication systems, and meters for transporting natural and/or supplemental gas from one point to another, usually from a point in or beyond the producing field or processing plant to another pipeline or to points of use.

Petroleum: A broadly defined class of liquid hydrocarbon mixtures. Included are crude oil, lease condensate, unfinished oils, refined products obtained from the processing of crude oil, and natural gas plant liquids.

Possible reserves: Those unproven reserves which analysis of geological and engineering data suggests are less likely than probable reserves to be commercially recoverable. Most companies assign a certainty value of 10 percent for possible reserves.

Price Band Mechanism: During its March 2000 meetings, OPEC adopted an informal price band mechanism whereby prices higher than $28 per barrel or lower than $22 per barrel for the OPEC basket would trigger automatic production adjustments. Prices sustained above the target range for twenty trading days were to result in an automatic production increase of 500,000 barrels per day (b/d). Prices below the target range for ten trading days were to result in cuts of 500,000 b/d.

Probable reserves: Those unproven reserves which analysis of geological and engineering data suggests are more likely than not to be commercially recoverable. Most companies assign a certainty value of 50 percent for probable reserves.

Proven reserves: Those quantities of petroleum which analysis of geological and engineering data indicates with reasonable certainty can be recovered in the future from known reservoirs, under existing economic and operating conditions. Most companies assign a certainty value of 90 percent for proven reserves.

Refinery (oil): An installation that manufactures finished fuels from oil, unfinished oils, natural gas liquids, other hydrocarbons, and alcohol.

Renewable energy: Energy obtained from sources that are essentially inexhaustible (unlike, for example, fossil fuels, of which there is a finite supply). Renewable sources of energy include conventional hydroelectric power, wood, waste, geothermal, wind, photovoltaic, and solar thermal energy.

Reservoir: A porous and permeable underground formation containing an individual and separate natural accumulation of producible hydrocarbons (crude oil and/or natural gas) that is confined by impermeable rock or water barriers and is characterized by a single natural pressure system.

"Seven Sisters": A phrase denoting the seven major oil companies that controlled most of the cheap Middle East oil between 1945 and 1973: Standard Oil Co. of New Jersey (later Exxon), Standard Oil Co. of New York (originally Socony; later, Mobil), Standard Oil Co. of California (Socal, later Chevron), Royal Dutch/Shell, Texaco, BP, and Gulf.

Spot price: The price for a onetime open market transaction for immediate delivery of the specific quantity of product at a specific location where the commodity is purchased "on the spot" at current market rates.

Stocks: Supplies of fuel or other energy source(s) stored for future use.

Well: A hole drilled in the earth for the purpose of (1) finding or producing crude oil or natural gas or (2) producing services related to the production of crude or natural gas.

Wellhead price: The price of oil or natural gas at the mouth of the well.

West Texas Intermediate (WTI): The "marker" crude in North America and the contract grade for the NYMEX crude oil future contract. It is widely accepted as the basis for pricing most American and Canadian crude oil.

Notes

Chapter 1. U.S. Energy Security

1. Edward L. Morse and Amy M. Jaffe, *Strategic Energy Policy Challenges for the Twenty-first Century*.
2. Energy Information Administration (EIA), *Energy in the United States: A Brief History and Current Trends*.
3. EIA, *Country Profile: USA*.
4. National Energy Policy Development Group (NEPDG), *National Energy Policy*.
5. EIA, *U.S. Natural Gas Markets: Recent Trends and Prospects for the Future*.
6. EIA, *Country Profile: USA*.
7. EIA, *Annual Energy Outlook*.
8. NEPDG, *National Energy Policy*.
9. Algeria was once the only source of LNG supply for the United States, but in recent years cargos have come from Trinidad and Tobago, Qatar, Nigeria, Australia, Oman, Indonesia, and the United Arab Emirates.
10. EIA, *Country Profile: USA*.
11. BP Amoco, *BP Statistical Review of World Energy*.
12. Robert J. Beck, "U.S. Drilling to Slump in 1999 Despite Oil, Gas Demand Gains."
13. NEPDG, *National Energy Policy*.
14. EIA, *Twenty-fifth Anniversary of the 1973 Oil Embargo*.
15. The term is borrowed from David A. Deese and Joseph S. Nye, *Energy and Security*.
16. For more details, see www.savedomesticoil.com.
17. Svante Karlsson, *Oil and the World Order: American Foreign Oil Policy*.
18. In 1995, with oil relatively abundant, Congress repealed the national speed limits and authorized states to set their own.
19. Edward W. Chester, *United States Oil Policy and Diplomacy*.
20. Seth P. Tillman, *The United States in the Middle East*.
21. EIA, *Potential Oil Production from the Coastal Plain of the Arctic National Wildlife Refuge: Updated Assessment*.
22. EIA, *Twenty-fifth Anniversary of the 1973 Oil Embargo*.
23. Paul Stevens, *The Economics of Energy*.
24. NEPDG, *National Energy Policy*.

25. EIA, *Annual Energy Outlook*.

26. In 2000, coal provided 52 percent of electricity generation and nuclear energy 20 percent, followed by natural gas with 16 percent. The rest was provided by hydropower, oil, and renewable sources.

27. In March 1979, a reactor at the Three Mile Island nuclear power plant in Harrisburg, Pennsylvania, suffered a partial meltdown. Within weeks, attorneys filed a class action suit against Metropolitan Edison Company (a subsidiary of General Public Utilities) on behalf of all businesses and residents within twenty-five miles of the plant. Over the next fifteen years, the case went to the Supreme Court and through various district and appeals courts. In June 1996, the lawsuit was finally dismissed.

28. EIA, *Country Profile: USA*.

29. The James A. Baker III Institute for Public Policy, *New Energy Technologies: A Policy Framework for Micronuclear Technology*.

30. U.S. Energy Department, *Strategic Petroleum Reserve*.

31. EIA, *Country Profile: USA*.

32. NEPDG, *National Energy Policy*.

33. Sarah Emerson, "SPR Drawdowns Trigger Law of Unintended Consequences."

34. EIA, *International Energy Outlook*.

35. Judith Gurney, "U.S. Faces Natural Gas Price Shock."

36. George Baker, "Mexico's New Energy Era: Pemex Development Tracking Fiscal, Technological Strategies."

37. BP Amoco, *BP Statistical Review of World Energy*.

38. EIA, *Persian Gulf Oil and Gas Exports Fact Sheet*.

Chapter 2. The Global Energy Scene

1. European Commission, *Green Paper: Towards a European Strategy for the Security of Energy Supply*.

2. EIA, *North Sea*.

3. David Young, "Politics or Technology."

4. Gerd Nonneman, "Saudi-European Relations, 1902–2001: A Pragmatic Quest for Relative Autonomy."

5. The deal was concluded after five years of Saudi attempts to buy F-15 fighter-bombers were defeated by effective opposition from the U.S. pro-Israeli lobby.

6. BP Amoco, *BP Statistical Review of World Energy*.

7. EIA, *Country Profile: Russia*.

8. EIA, *Russia: Energy Sector Restructuring*.

9. EIA, *Country Profile: Russia*.

10. Isabel Gorst, "Change Looms over the Mighty Gazprom."

11. The Energy Charter Treaty set out a sectoral legal framework for energy. Comprising some fifty articles, it deals with issues pertaining to investment, trade, transit, competition, and environmental concerns. For the text of the treaty and more details, see the organization's website at www.encharter.org.

12. Ria Kemper, "EU Looks to Secure Russian Supply Future."
13. Valeria Korchagina, "Russia Gears Up to Export More Oil with BPS."
14. This proposed pipeline is designed to bypass Ukraine, which has been repeatedly accused by Russian officials of illegally siphoning Russian gas.
15. EIA, *Oil and Gas Export Pipelines*.
16. The proposed pipeline would supply China with 400,000 barrels a day starting in 2005, the equivalent by then of 26 percent of China's projected net imports.
17. Alan Troner, "Russian Far East Natural Gas."
18. In October 2001, ExxonMobil announced that a consortium it leads will spend $4 billion over five years to develop large offshore oil and natural gas fields in Russia's Far Eastern Sakhalin region—the single largest foreign investment in Russia.
19. Igor Khripunov and Mary M. Matthews, "Russia's Oil and Gas Interest Group and Its Foreign Policy Agenda."
20. Naturally, there is a broad variation among individual states.
21. EIA, *East Asia: The Energy Situation*.
22. Edward L. Morse, "The Tigers Are Roaring Back . . . But Their Stripes Have Changed."
23. In 1998, Pacific Asia imported 4.6 million barrels per day from the Persian Gulf. The figure is projected to jump to 13.5 million barrels per day by 2020. See EIA, *International Energy Outlook 2001*.
24. EIA, *Country Profile: China*.
25. Neil Heywood, "Farewell to Self-Sufficiency."
26. EIA, *Country Profile: China*.
27. Robert A. Manning, "The Asian Energy Predicament."
28. Fred R. von der Mehden, *Japan's Relations with Primary Energy Suppliers*.
29. Baker Institute for Public Policy, *Japanese Energy Security and Changing Global Energy Markets: An Analysis of Northeast Asian Energy Cooperation and Japan's Evolving Leadership Role in the Region*.
30. EIA, *Country Profile: Japan*.
31. International Emissions Trading Association, *What Is the Science behind Climate Change?*
32. The Kyoto Protocol will enter into force upon ratification by a mimimum of fifty-five parties that accounted for at least 55 percent of the Annex I carbon dioxide emissions in 1990. The discussion in this section is based on the text of the Kyoto Protocol, on-line at www.unfccc.org.
33. The six greenhouse gases specified in the Kyoto Protocol are carbon dioxide, methane, nitrous oxide, hydrofluorocarbons, perfluorocarbons, and sulphur hexafluoride. The emission reduction targets for 2008–12 (with respect to 1990 levels) taken on by industrialized countries under Kyoto Protocol are Switzerland, Central and East European states, European Union, –8 percent; United States, –7 percent; Canada, Hungary, Japan, and Poland, –6 percent; New Zealand, Russia,

and Ukraine, 0 percent; Norway, +1 percent; Australia, +8 percent; Iceland, +10 percent.

34. These last three options (joint implementation, clean development mechanism, and emissions trading) are known collectively as flexibility mechanisms.

35. Peter Davies, "Climate Change: What If We Can't Unravel the Gordian Knot?"

36. "The Kyoto Protocol and the U.S. Came, Too."

37. Environmentalists often point out that, with only 5 percent of the world's people, the United States emits 25 percent of the world's greenhouse gases. What is missing from the equation, however, is that the United States also produces about 25 percent of the world's goods and services.

38. Thomas G. Burns, "Post-Kyoto Climate Policy Possible."

39. EIA, *Impacts of the Kyoto Protocol on U.S. Energy Markets and Economic Activity.*

40. Shokri Ghanem, Rezki Lounnas, and Garry Brennand, "The Impact of Emissions Trading on OPEC."

41. "Saudi Arabia on Kyoto Protocol Developments."

Chapter 3. Managing Dependence: American-Saudi Oil Diplomacy

1. John Marlowe, *The Persian Gulf in the Twentieth Century.*

2. George Lenczowski, *Oil and State in the Middle East.* In addition to these four companies, the Gulf Oil Corporation, the Royal Dutch/Shell, and the Anglo-Persian Oil Company (later Anglo-Iranian and then British Petroleum) dominated the oil industry, particularly in the Middle East, until the 1970s. These companies are known as the Seven Sisters.

3. Keith McLachlan, "Oil in the Persian Gulf Area."

4. For more details, see Aramco's website at www.saudiaramco.com.

5. Given the crucial role oil plays in the global economy, some analysts called it a hydrocarbon economy.

6. Peter E. Jones, *Oil: A Practical Guide to the Economics of World Petroleum.*

7. Paul McDonald, "What Is the Price of Oil?"

8. EIA, *OPEC Revenues Fact Sheet*, February 2001.

9. Iran also changed to the "B-Wave" in early 2001.

10. A. M. Samsam Bakhtiari, "The Price of Crude Oil."

11. Joe Stork, *Middle East Oil and the Energy Crisis.*

12. Seth P. Tillman, *The United States in the Middle East.*

13. The embargo was also imposed on the Netherlands, Portugal, Rhodesia, and South Africa.

14. In an interview, Kissinger said, "I am not saying that there's no circumstance where we would not use force." See "Kissinger on Oil, Food, and Trade."

15. France demurred on the theory that the agency would be confrontational, while it favored good relations with oil producers and the Arab world. In 1992, France dropped its objections and joined the agency.

16. "Assuring Energy Security."

17. Alan Richards, "Oil Wealth in the Arab World: Whence, to Whom, and Whither?"

18. After signing a peace treaty with Israel, Egypt was expelled from the Arab League, which then moved its headquarters from Cairo to Tunis.

19. EIA, *Country Profile: Saudi Arabia*.

20. Ali al-Naimi, "Saudi Oil Policy Combines Stability with Strength, Looks for Diversity."

21. In 1999, the Center for Global Energy Studies estimated the cost to Saudi Arabia of maintaining its idle capacity at $500 million each year. See Ahmed Zaki Yamani, "OPEC Should Take Long-Term Approach to Balancing Oil Supply-Demand Equation."

22. In March 2001, Energy Secretary Spencer Abraham noted that the country's past three recessions were all tied to rising energy prices and that "there is strong evidence that the latest crisis is already having a negative effect." See Eric Pianin, "U.S. Faces an Energy Shortfall, Bush Says."

23. EIA, *OPEC Revenues Fact Sheet*, May 1998.

24. "Energy Market Report."

25. EIA, *International Petroleum Monthly*.

26. Tom Manning, "Asian Financial Crisis to Slow Growth in Global Oil Demand."

27. "Editorial: Oil Prices and Layoffs."

28. "U.S. Crude Reserves Plunged 7 Percent in 1998."

29. "OPEC Ministers Approve Huge 1.7 Million B/D Production Cuts."

30. "No Inflation Fear."

31. National Energy Policy Development Group, *National Energy Policy*.

32. EIA, *OPEC Revenues Fact Sheet*, March 2001.

33. "Saudi Warning of Further Export Hike Trims $2/B Off Crude Prices."

34. "Producers Agree to Maintain Production Cutbacks until End-March; Positions Differ on What to Do Next."

35. For the details of this agreement, see Peter Behr, "Eleven Firms to Get Oil from Reserve," and Neela Banerjee, "Eleven Companies Buy Crude Oil from Reserves."

36. George W. Bush, then running for president, said the move was purely political.

37. EIA, *OPEC Revenues Fact Sheet*, February 2001.

38. Michael C. Lynch, "Oil Prices Enter a New Era."

39. Those attending the meeting with Abdullah included senior executives from the four American oil giants—Mobil, Exxon, Texaco, and Chevron—that established the Arabian American Oil Company, now known as Saudi Aramco, in the 1930s. Senior executives from Atlantic Richfield, Conoco, and Phillips Petroleum also attended.

40. David B. Ottaway and Martha M. Hamilton, "Saudis Talk with Seven U.S. Oil Firms."
41. Edward L. Morse, "A New Political Economy of Oil?"
42. In 2000, Saudi Arabia's share of U.S. crude oil imports was 17 percent.
43. EIA, *Country Profile: Saudi Arabia*.
44. James Gavin, "The Walls Come Down."
45. "Great Expectations."
46. "Negotiations for Saudi Natural Gas Initiative to Start in Second Half 2000."
47. EIA, *Country Profile: Saudi Arabia*.
48. For more details, see Saudi Aramco website at www.saudiaramco.com.
49. Prince Faisal Ibn Turki Ibn Abd al-Aziz, "Perspectives on the Saudi Energy Industry."
50. EIA, *Country Profile: Saudi Arabia*.
51. The list includes BP Amoco, Chevron, Conoco, ENI, ExxonMobil, Marathon, Enron-Occidental, Phillips Petroleum, Royal Dutch Shell Group, Texaco, and TotalFinaElf.
52. Ian Seymour, "Opportunities for Upstream Investment in the Middle East by IOCs."
53. Robin Allen, "ExxonMobil Seals Saudi Deal," *Financial Times,* June 4, 2001.
54. TotalFinaElf, BP PLC, Conoco, Phillips Petroleum, and Occidental Petroleum were selected as junior partners in the three projects.
55. The National Energy Policy Development Group, headed by Vice President Dick Cheney, stressed that promoting American investment will be a core element of U.S. engagement with major foreign oil producers. See NEPDG, *National Energy Policy*.
56. Saudi Arabia's spare capacity is estimated between 2 and 3 million barrels per day.
57. Research done in the late 1990s by the Center for Global Energy Studies suggests that implementation of the Kyoto Protocol would require by 2010 around 4.5 million b/d less oil consumption in the world than otherwise would have been the case.
58. David Buchan, "Saudis Attack Anti-Oil Policies."
59. For details, see www.energyforum.gov.sa.
60. Cited in George W. Stocking, *Middle East Oil: A Study in Political and Economic Controversy.*
61. George Lenczowski, *American Presidents and the Middle East.*
62. In January 2002, Senator Carl M. Levin (D-Mich.), chairman of the Armed Services Committee, said that the United States should consider moving its forces out of the kingdom and find a place that is more hospitable.
63. David Butter, "Special Report—Saudi Arabia: Taking a Lead."

64. Previously, foreign companies were limited to a 49 percent share and obliged to form a joint enterprises with either state or private Saudi firms.

65. David B. Ottaway, "U.S. Oil CEOs and Saudis to Meet."

66. Howard Schneider, "Bombing in Saudi City Kills American."

Chapter 4. The United States and Iraq: Continuity and Change

1. Paul A. Gigot, "A Great American Screw-Up: The U.S. and Iraq, 1980–1990."

2. Zachary Karabell, "Backfire: U.S. Policy toward Iraq, 1988–2 August 1990."

3. "Bush Says He Erred in Assuming Hussein Would Fall after War."

4. EIA, *Country Profile: Iraq*.

5. Keith McLachlan, "Oil in the Persian Gulf Area."

6. Edward W. Chester, *United States Oil Policy and Diplomacy*.

7. The five companies are Standard Oil of New Jersey, Standard Oil of New York, Gulf Refining Company, Atlantic Refining Company, and Pan American Petroleum and Transport Company. Together they held 23.75 percent of the shares in the IPC. See Svante Karlsson, *Oil and the World Order: American Foreign Oil Policy*.

8. "Iraq: Sanctions Policy Reaches Fork in the Road."

9. Kurt S. Abraham, "What Is the Status of Iraq's Upstream Sector?"

10. Colum Lynch, "U.N. Chief Faults U.S., Britain for Iraqi Supply Delays."

11. EIA, *Country Profile: Iraq*.

12. George Lenczowski, "Major Pipelines in the Middle East: Problems and Prospects."

13. "Saudi Arabia Seizes IPSA Pipeline."

14. As of early 2001, Iraqi officials were claiming that the pipeline had been rehabilitated.

15. As mentioned, the ceiling on how much oil Iraq can sell was lifted by Security Council Resolution 1284 in December 1999.

16. Abbas Alnasrawi, "Iraq: Economic Sanctions and Consequences, 1990–2000."

17. United Nations Development Program, *Human Development Report*.

18. Denis Halliday, "Iraq and the UN's Weapons of Mass Destruction." See also Halliday, "The Impact of the UN Sanctions on the People of Iraq."

19. Sarah Graham-Brown, "Sanctioning Iraq: A Failed Policy."

20. Daniel Byman, "After the Storm: U.S. Policy toward Iraq since 1991."

21. Nicole Winfield, "UN: Euro Account for Iraq Feasible."

22. Alan Sipress, "More Iraqi Oil Evading Sanctions."

23. William Drozdiak, "Iraq Imperils OPEC Plan."

24. "More Trouble from Saddam."

25. Marion Farouk-Sluglett and Peter Sluglett, *Iraq since 1958: From Revolution to Dictatorship*.

26. Helen Avati, "Total Targeting Low-Cost Reserves."
27. International Institute for Strategic Studies, *Strategic Survey*.
28. "U.S./UN Begin Review of Iraqi Sanctions."
29. "Iraq: Russia Gets $40 Billion in Contracts."
30. Fadhil J. Chalabi, "The Opening of Iraq: Post-Sanctions Iraqi Oil, Its Effects on World Oil Prices."
31. "Iraq Changes Terms of Future Oil Project Contracts."
32. At the time of the Persian Gulf war, Colin Powell was chairman of the Joint Chiefs of Staff and Dick Cheney was Secretary of Defense.
33. The full text of the confirmation hearing is posted on the State Department's website at www.state.gov.
34. Howard Schneider, "Review of Iraq Sanctions Reflects State of Disarray."
35. "Security Council Renews Oil-for-Food Program, Pledges to Revise Iraqi Sanctions Next June."
36. John Ward Anderson, "Iraq's Neighbors Feel Pain of Sanctions."
37. For example, see Shibley Telhami, "Time for Realism on Handling Iraq."
38. Avner Yaniv, "Israel Faces Iraq: The Politics of Confrontation."
39. Ibrahim Ibrahim, *The Gulf Crisis: Background and Consequences*.
40. For a thorough analysis of this raid, see Shai Feldman, "The Bombing of Osiraq—Revisited," and Jed C. Snyder, "The Road to Osiraq: Baghdad's Quest for the Bomb."
41. For a recent comprehensive study of Israel's nuclear program, see Avner Cohen, *Israel and the Bomb*.
42. Ahmed Hashim, "Iraq: Profile of a Nuclear Addict."
43. Anthony H. Cordesman, "The Changing Military Balance in the Gulf."
44. Hashim, "Iraq: Profile of a Nuclear Addict."
45. Gawdat Bahgat, "An Overview of Gulf Security: Oil and Weapons of Mass Destruction."
46. David Albright and Khidhir Hamza, "Iraq's Reconstitution of Its Nuclear Weapons Program."
47. David A. Kay, "Denial and Deception Practices of WMD Proliferators: Iraq and Beyond."
48. David Malone, "Good-bye, UNSCOM: A Sorry Tale in U.S.–UN Relations."
49. Youssef M. Ibrahim, "Higher Hopes in Baghdad for Ending UN Embargo."
50. Tim Weiner, "U.S. Spied on Iraq under UN Cover, Officials Now Say."
51. Barton Gellman, "Annan Suspicious of UNSCOM Role."
52. Gellman, "Israel Gave Key Help to UN Team in Iraq."
53. Judith Miller, "UN Panel Urges New Inspections in Iraq."
54. Colum Lynch and John Lancaster, "UN Votes to Renew Iraq Inspections."
55. Meghan L. O'Sullivan, *Iraq: Time for a Modified Approach*.
56. A detailed assessment of Iraq's WMD capabilities can be found in the U.S. Defense Department, *Proliferation: Threat and Response*.

57. This assessment is drawn from a statement by Hans Blix, chairman of the United Nations Monitoring, Verification, and Inspection Commission. See Barbara Crossette, "Iraq Suspected of Secret Germ War Effort," and Marc Weller, "The U.S., Iraq, and the Use of Force in Unipolar World."

58. Crossette, "Pressing for Iraqi's Overthrow, U.S. Appeals for Arab Support."

59. Thomas L. McNaugher, "Arms Sales and Arms Embargoes in the Persian Gulf: The Real Dilemmas of Dual Containment."

60. Carl Hulse, "U.S. Said to Approve Anti-Hussein Funding."

61. Philip Shenon, "U.S. General Warns of Dangers in Trying to Topple Iraqi."

62. Daniel Byman, Kenneth Pollack, and Gideon Rose, "The Rollback Fantasy."

63. These include former presidential candidate Senator John McCain (R-Ariz.); former vice presidential candidate Senator Joseph I. Lieberman (D-Conn.); the ranking Republican on the Senate Foreign Relations Committee, at the time, Jesse Helms (N.C.); Senate Minority Leader, at the time, Trent Lott (R-Miss.); House International Relations Committee Chairman Henry J. Hyde (R-Ill.); and the ranking Republican on the Senate Intelligence Committee, Richard C. Shelby (Ala.).

64. "Ten Leading Lawmakers Urge Targeting of Iraq."

65. Nina J. Easton, "The Hawk."

66. Henry A. Kissinger, "Phase II and Iraq."

67. National Security Adviser Condoleezza Rice is believed to be not quite in either camp.

68. Michael Dobbs, "Old Strategy on Iraq Sparks New Debate."

69. A. M. Samsam Bakhtiari, "OPEC Capacity Potential Needed to Meet Projected Demand Not Likely to Materialize."

Chapter 5. The United States and Iran: Prospects for Rapprochement

1. George Lenczowski, *Oil and State in the Middle East.*
2. Colin J. Campbell, *The Coming Oil Crisis.*
3. Keith McLachlan, "Oil in the Persian Gulf Area."
4. Lenczowski, *Oil and State in the Middle East.*
5. Edward W. Chester, *United States Oil Policy and Diplomacy: A Twentieth-Century Overview.*
6. Vahe Petrossian, "Special Report Iran: Year of Decision."
7. Petrossian, "Raising the Stakes."
8. Derek Brower, "Explorers in Defiant Mood."
9. BP Amoco, *BP Statistical Review of World Energy.*
10. Manouchehr Takin, "Iranian Gas to Europe?"
11. BP Amoco, *BP Statistical Review of World Energy.*
12. A. M. Samsam Bakhtiari and F. Shahbudaghlou, "Energy Consumption in the Islamic Republic of Iran."

13. EIA, *Country Profile: Iran.*
14. "Tehran Seeks More Gas Buy-Backs."
15. Petrossian, "Special Report Gas: Aiming East."
16. "Enterprise Leads Way in South Pars."
17. "Iran's NIOC Discovers Supergiant Gas, Condensate Field."
18. "Iran Reveals Second Big Gas Discovery."
19. Angus McDowall, "Special Report Oil and Gas: Taking Stock."
20. "Studies Start for IGAT-4."
21. International Energy Agency, *Caspian Oil and Gas.*
22. For more details, see Emre Engur, "Turkey Determined to Remain at the Center of East-West Energy Corridor."
23. EIA, *Country Profile: Turkey.*
24. EIA, *Country Profile: Iran.*
25. Toby Shelley, "BG in LNG Venture with Iran."
26. An initial accord on an Iranian-Armenian pipeline was signed in 1992 and modified in 1995 but was never implemented due to financial problems. In late 2001, the two countries renewed their interest in building a pipeline.
27. "TotalFinaElf Invites Bids for Doroud Onshore Package in Iran."
28. Ian Seymour, "Opportunities for Upstream Investment in the Middle East by International Oil Companies."
29. The consortium comprises the UK's BG International and Lasmo, France's TotalFinaElf and Gaz de France, UK-based BP Amoco, Agip of Italy, the Royal Dutch/Shell Group, Australia's BHP Petroleum, and Petronas of Malaysia.
30. The name *Azadegan* (the Freed) was chosen in reference to the tens of thousands of former Iranian prisoners of war from the war with Iraq.
31. Dean E. Gaddy, "Iran Expands Middle East Influence."
32. International Energy Agency, *Caspian Oil and Gas.*
33. Seymour, "Kuwait's Upstream Oil Opening in the Context of Parallel Development Elsewhere."
34. Kirsten Bindemann, "A Little Bit of Opening Up: The Middle East Invites Bids by Foreign Oil Companies."
35. Fereidun Fesharaki and Mehdi Varzi, "Investment Opportunities Starting to Open Up in Iran's Petroleum Sector."
36. "Iran's Expediency Council Rescues Buybacks as Contracts Spending Rolls Over into New Budget Year."
37. "United States person" is broadly defined to include U.S. citizens, U.S. permanent resident aliens, entities organized under U.S. law, or any person located in the United States.
38. "U.S. Sanctions: Fresh Thinking on Capitol Hill."
39. The entire text of Executive Order 13059 is available on-line at www.iraniantrade.org/13059.htm.
40. EIA, *World Energy Sanctions.*
41. Gawdat Bahgat, "U.S.–Iranian Relations: Sanctions and the Caspian Sea."

42. Elaine Sciolino, "Iranians Back Their Leader's Volatile Experiment."
43. "U.S. Reviews Iran and Sanctions Policies."
44. Those opposing were Charles Hagel (R-Neb.) and Richard Lugar (R-Ind.).
45. Those opposing were Ron Paul (R-Tex.), Earl Hilliard (D-Ala.), John Conyers (D-Mich.), John LaFalce (D-N.Y.), Cynthia McKinney (D-Ga.), and Nick Rahall (D-W.Va.).
46. In the late 1990s, commercial quantities of natural gas were discovered off the coast of Israel.
47. For a discussion of Israel's oil needs and policies, see Stephen H. Longrigg, *Oil in the Middle East*.
48. Cited in M. G. Weinbaum, "Iran and Israel: The Discrete Entente."
49. Harvey Morris, "Israel Fears Threat of Iran Links with Palestinians."
50. In May 2000, Israeli forces withdrew from the so-called security zone. However, Hizbollah still attacks Israeli troops in a 100-square-mile area known as Shebaa Farms, which it claims as part of Lebanon.
51. Aluf Benn, "Israel Once Again Sees Iran as a Cause for Concern."
52. Lee Hochstrader, "Iran Implicated by Captain in Seized Weapons Shipment."
53. "Interview with President Khatami."
54. Herb Keinon, "Iran Using PA as Proxy against Israel-Senior Official."
55. "Palestinian-Iran Ties Worry Israel."
56. Brian Whitaker, "Voyage of the Arms Ship," and Miriam Shaviv, "A Chill Wind from Tehran."
57. Gil Sedan, "Iran Has Long Anti-Israel History."
58. Amir Oren, "Israel Asks America to Strike Western Iraq First, If It Decides to Fight Saddam."
59. U.S. State Department, *Patterns of Global Terrorism, 2000*.
60. R. James Woolsey, "Appeasement Will Only Encourage Iran."
61. Jeffrey Smith, "Khatami Wants to End Terrorism, Officials Say."
62. Molly Moore, "Mistrust Still Divides Iran and U.S."
63. "Interview with President Khatami."
64. U.S. Defense Department, *Proliferation: Threat and Response*.
65. Shahram Chubin, "Iran's Strategic Environment and Nuclear Weapons."
66. Daniel Byman, Shahram Chubin, Anoushiravan Ehteshami, and Jerrold Green, *Iran's Security Policy in the Post-Revolutionary Era*.
67. Muhammad Javad Zarif and Muhammad Reza Alborzi, "Weapons of Mass Destruction in Iran's Security Paradigm: The Case of Chemical Weapons."
68. Geoffrey Kemp, "Iran: Can the United States Do a Deal?"
69. Seymour M. Hersh, "Analysis of National Security: The Iran Game."
70. Siamak Namazi, "U.S.–Iran Relations after 11 September."
71. "Rafsanjani Says U.S. Aiming to 'Besiege' Iran, Warns Washington, Europe."
72. Edward Cody, "Iran Said to Aid Afghan Commander."

73. These are Senators Arlen Specter (R-Penn.) and Mike DeWine (R-Ohio) and Representatives Bob Ney (R-Ohio), Jim Leach (R-Iowa), Sheila Jackson Lee (D-Tex.), and Paul Kanjorski (D-Penn.).
74. Janine Zacharia, "World Bank Set to Consider Iran Loans."
75. Economist Intelligence Unit, *Country Analysis: Iran*.

Chapter 6. The Geopolitics of the Caspian Sea

1. Some historians claim that commercial oil production started near Baku in 1848. See Colin J. Campbell, *The Coming Oil Crisis*.
2. Rosemarie Forsythe, *The Politics of Oil in the Caucasus and Central Asia*.
3. International Energy Agency, *Caspian Oil and Gas*.
4. EIA, *Country Profile: Kazakhstan*.
5. EIA, *Country Profile: Azerbaijan*.
6. Julia Nanay, "The U.S. in the Caspian: The Divergence of Political and Commercial Interests."
7. International Institute for Strategic Studies, *Strategic Survey*.
8. Paul Horsnell, "Caspian Oil and Gas: A Game, If Not a Great Game."
9. It is important to point out that there is no agreement on how to divide the Caspian Sea among the five littoral states.
10. "Caspian Reserves Assessed."
11. "Russia Moves Ahead with Caspian Petroleum Projects."
12. "World News: More Doubts over Azerbaijan Energy."
13. EIA, *Country Profile: Azerbaijan*.
14. OKIOC consists of Philips Petroleum and ExxonMobil of the United States, Italy's Agip, Britain's BG, BP Amoco, and Royal Dutch/Shell, Japan's Inpex, TotalFina of France, and Norway's Statoil.
15. EIA, *Country Profile: Kazakhstan*.
16. Christopher Pala, "AGIP of Italy Picked to Manage Big Kazakh Oil Field."
17. Hilary McCutcheon, "Caspian Production Potential: Discoveries Alter Caspian Region Energy Potential."
18. In the nineteenth century, the rivalry between the British and Russian empires over roads and trade to Asia was called the "Great Game."
19. CIS includes former Soviet republics.
20. The members in Partnership for Peace are Albania, Armenia, Austria, Azerbaijan, Belarus, Bulgaria, Czech Republic, Estonia, Finland, Georgia, Hungary, Kazakhstan, Kyrgyzstan, Latvia, Lithuania, Malta, Moldova, Poland, Romania, Russia, Slovakia, Slovenia, Sweden, the former Yugoslav Republic of Macedonia, Turkmenistan, Ukraine, and Uzbekistan.
21. The Black Sea Economic Cooperation Pact comprises Russia, Ukraine, Turkey, Georgia, Romania, Bulgaria, Albania, Armenia, Azerbaijan, Moldova, and Greece.
22. The group brings together China, Russia, Kazakhstan, Kyrgyzstan, Tajikistan, and Uzbekistan.

23. GUUAM comprises Georgia, Uzbekistan, Ukraine, Azerbaijan, and Moldova.

24. The ECO was established in 1985 by Iran, Pakistan, and Turkey. In 1992, the organization was expanded to include seven new members: Afghanistan, Azerbaijan, Kazakhstan, Kyrgyzstan, Tajikistan, Turkmenistan, and Uzbekistan.

25. International Energy Agency, *Caspian Oil and Gas.*

26. James A. Baker III Institute for Public Policy, *Running on Empty: Prospects for Future World Oil Supplies.*

27. Carol R. Saivetz, "Caspian Geopolitics: The View from Moscow."

28. Edmund Herzig, "Iran and Central Asia."

29. Iran is not a member of this regional organization, which was created in 1981 and comprises Bahrain, Kuwait, Oman, Qatar, Saudi Arabia, and the United Arab Emirates.

30. "Iran Offers Discounts to Support Caspian Export Offer."

31. Paul Stevens, "Consumer Governments, Energy Security of Supply, and the Aftermath of 11 September."

32. Stanislav Cherniavskii, "Current Problems of the Transcaucasus."

33. Fred Halliday, "Condemned to React, Unable to Influence: Iran and Transcaucasia."

34. Shireen T. Hunter, *The Transcaucasus in Transition: Nation-Building and Conflict.*

35. EIA, *Caspian Regional Conflict.*

36. Brenda Shaffer, *U.S.–Russian Relations: Implications for the Caspian Region.*

37. Hunter, *The Transcaucasus in Transition.*

38. Roland Sinker, "The Management of a Transboundary Energy Resource: The Oil and Gas of the Caspian Sea."

39. International Energy Agency, *Caspian Oil and Gas.*

40. Theodore C. Jonas, "Parting the Sea: Caspian Littoral States Seek Boundary Disputes' Resolution."

41. Iurii Merzliakov, "Legal Status of the Caspian Sea."

42. At the time of this writing (late 2001), this agreement has not been ratified by the parliaments of either country.

43. Michael Lelyveld, "Russian, Iranian Presidents Fail to Agree on Caspian Sea Division."

44. The Caspian Sea contains about 90 percent of the world's sturgeon.

45. According to the median line theory, the sea should be divided based on the median line of the sea, which has the same distance from both shores.

46. These figures were cited by Abbas Maleki, chairman of the International Institute for Caspian Studies, a Tehran-based think tank, on March 22, 2001, online at www.payvand.com.

47. For a detailed discussion of these routes, see Ronald Soligo and Amy Jaffe, *The Economics of Pipeline Routes: The Conundrum of Oil Exports from the*

Caspian Basin, and Geoffrey Kemp and Robert E. Harkavy, *Strategic Geography and the Changing Middle East*.

48. EIA, *Caspian Tables and Maps*.

49. Ferruh Demirmen, "Despite Recent Gains in Momentum, Prospects for the Baku-Ceyhan Caspian Oil Export Line Remain Doubtful."

50. EIA, *Country Profile: Azerbaijan*, May 2001.

51. On September 29, 2001, Presidents Heydar Aliyev of Azerbaijan and Eduard Shevardnadze of Georgia signed an agreement regarding the construction of the $1 billion pipeline across Georgia from Azerbaijan to Turkey.

52. Robert Smith, "Politics, Production Levels to Determine Caspian Area Energy Export Options."

53. Douglas Frantz, "Russia's New Reach: Gas Pipeline to Turkey."

54. "Gas Exports to Rise as New Lines Come on Stream."

55. In 2001, TengizChevroil consisted of Chevron 50 percent, ExxonMobil 25 percent, Kazakhstan 20 percent, and LUKARCO 5 percent.

56. Shareholders of the CPC include Russian, American, British, Kazak, and Omani companies.

57. Dean E. Gaddy, "Iran Expands Middle East Influence."

58. Several European, particularly British, companies have been involved in swap arrangements with Iran since the late 1990s, including Monument Resources Petroleum (later to become Lasmo Oil), Dragon Oil, Burren Energy, and Larmag Energy Associates.

Bibliography

Abadi, Jacob. "Israel's Relations with Oman and the Persian Gulf States." *Journal of South Asian and Middle Eastern Studies* 20, no. 1 (Fall 1996): 46–73.

Abraham, Kurt S. "What Is the Status of Iraq's Upstream Sector?" *Journal of Energy and Development* 24, no. 2 (Spring 2000): 203–14.

Albright, David, and Khidhir Hamza. "Iraq's Reconstitution of Its Nuclear Weapons Program." *Arms Control Today* 28, no. 7 (October 1998): 9–15.

Allen, Robin. "ExxonMobil Seals Saudi Deal." *Financial Times*, June 4, 2001.

Allison, Roy, and Lena Jonson, eds. *Central Asian Security: The New International Context*. Washington, D.C.: Brookings Institution, 2001.

Alnasrawi, Abbas. "Iraq: Economic Sanctions and Consequences, 1990–2000." *Third World Quarterly* 22, no. 2 (April 2001): 205–18.

Amirahmadi, Hooshang. "Iran's Development: Evaluation and Challenges." *Third World Quarterly* 17, no. 1 (March 1998): 123–47.

Amuzegar, Jahangir. *Iran's Economy under the Islamic Republic*. London: I. B. Tauris, 1993.

———. "Iran's Virtual Democracy at a Turning Point." *SAIS Review* 6, no. 2 (Summer/Fall 2000): 93–109.

Anderson, John Ward. "Iraq's Neighbors Feel Pain of Sanctions." *Washington Post*, July 1, 2001.

"Assuring Energy Security." *Energy* 25, no. 4 (Fall 2000): 4–10.

Avati, Helen. "Total Targeting Low-Cost Reserves." *Petroleum Economist* 66, no. 2 (February 1999): 32–33.

al-Aziz, Prince Faisal Ibn Turki Ibn Abd. "Perspectives on the Saudi Energy Industry." *Middle East Economic Survey* 44, no. 10 (March 5, 2001), on-line at www.mees.com.

Bahgat, Gawdat. "An Overview of Gulf Security: Oil and Weapons of Mass Destruction." *Disarmament Diplomacy*, no. 33 (December 1998–January 1999): 6–11.

———. "U.S.–Iranian Relations: Sanctions and the Caspian Sea." *Security Dialogue* 32, no. 2 (June 2001): 231–43.

Baker, George. "Mexico's New Energy Era: Pemex Development Tracking Fiscal, Technological Strategies." *Oil and Gas Journal* 99, no. 9 (May 2001): 58–74.

Baker (James A. III) Institute for Public Policy. *Japanese Energy Security and Changing Global Energy Markets: An Analysis of Northeast Asian Energy*

Cooperation and Japan's Evolving Leadership Role in the Region. Baker Institute Study no. 13. Houston: Rice University, May 2000.

———. *New Energy Technologies: A Policy Framework for Micronuclear Technology.* Baker Institute Study no. 17. Houston: Rice University, October 2001.

———. *Running on Empty: Prospects for Future World Oil Supplies.* Houston: Rice University, 2000.

Bakhtiari, A. M., and F. Shahbudaghlou. "Energy Consumption in the Islamic Republic of Iran." *OPEC Review* 24, no. 3 (September 2000): 211–29.

Baktiari, Bahman. *Parliamentary Politics in Revolutionary Iran.* Gainesville: University Press of Florida, 1996.

Baktiari, Bahman, and F. Shahbudaghlou. "Energy Consumption in the Islamic Republic of Iran." *OPEC Review* 24, no. 3 (September 2000): 211–29.

Bakhtiari, Samsam A. M. "OPEC Capacity Potential Needed to Meet Projected Demand Not Likely to Materialize." *Oil and Gas Journal* 99, no. 28 (July 9, 2001): 69.

———. "The Price of Crude Oil." *OPEC Review* 23, no. 1 (March 1999): 1–21.

Banerjee, Neela. "Eleven Companies Buy Crude Oil from Reserves." *New York Times*, October 5, 2000.

Baram, Amatzia, and Barry Rubin, eds. *Iraq's Road to War.* New York: St. Martin's Press, 1993.

Beck, Robert J. "U.S. Drilling to Slump in 1999 Despite Oil, Gas Demand Gains." *Oil and Gas Journal* 97, no. 4 (January 25, 1996): 60.

Behr, Peter. "Eleven Firms to Get Oil from Reserve." *Washington Post*, October 5, 2000.

Benn, Aluf. "Israel Once Again Sees Iran as a Cause for Concern." *Ha'aretz*, May 7, 2001.

Bialer, Uri. "The Iranian Connection in Israel's Foreign Policy." *Middle East Journal* 39, no. 2 (spring 1985): 292–315.

Bindemann, Kirsten. "A Little Bit of Opening Up: The Middle East Invites Bids by Foreign Oil Companies." *Oxford Institute for Energy Studies Monthly Comment*, December 1999, on-line at www.oxfordenergy.org.

Blackwill, Robert, and Michael Sturmer, eds. *Allies Divided: Transatlantic Policies for the Greater Middle East.* Cambridge: MIT Press, 1997.

BP Amoco. *BP Statistical Review of World Energy.* London: BP Amoco, 2001.

Brower, Derek. "Explorers in Defiant Mood." *Petroleum Economist* 66, no. 3 (March 1999): 22–25.

Buchan, David. "Saudis Attack Anti-Oil Policies." *Financial Times*, April 25, 2001.

Burns, Thomas G. "Post-Kyoto Climate Policy Possible." *Oil and Gas Journal* 99, no. 30 (July 23, 2001): 20–26.

"Bush Says He Erred in Assuming Hussein Would Fall after War." *New York Times*, January 15, 1996.

Butter, David. "Special Report—Saudi Arabia: Taking a Lead." *Middle East Economic Digest* 44, no. 48 (December 2000): 25–44.

Byman, Daniel. "After the Storm: U.S. Policy toward Iraq since 1991." *Political Science Quarterly* 115, no. 4 (Winter 2000-2001): 493–516.

Byman, Daniel, Kenneth Pollack, and Gideon Rose. "The Rollback Fantasy." *Foreign Affairs* 78, no. 1 (January/February 1999): 24–41.

Byman, Daniel, Shahram Chubin, Anoushiravan Ehteshami, and Jerrold Green. *Iran's Security Policy in the Post-Revolutionary Era*. Santa Monica: RAND, 2001.

Campbell, Colin J. *The Coming Oil Crisis*. Essex: Multi-Science, 1997.

"Caspian Reserves Assessed." *Financial Times*, September 19, 2000.

Central Intelligence Agency. *Handbook of International Economic Statistics*. Washington, D.C.: Government Printing Office, 1996.

Chalabi, Fadhil J. "The Opening of Iraq: Post-Sanctions Iraqi Oil, Its Effects on World Oil Prices." *Oil and Gas Journal* 98, no. 7 (February 14, 2000): 41–43.

Cherniavskii, Stanislav. "Current Problems of the Transcaucasus." *International Affairs* 45, no. 5 (October 1999): 137–46.

Chester, Edward W. *United States Oil Policy and Diplomacy: A Twentieth-Century Overview*. Westport, Conn.: Greenwood Press, 1983.

Chubin, Shahram. "Iran's Strategic Environment and Nuclear Weapons." In *Iran's Nuclear Weapons Options: Issues and Analysis,* ed. Geoffrey Kemp, 17–33. Washington, D.C.: Nixon Center, 2001.

Cody, Edward. "Iran Said to Aid Afghan Commander." *Washington Post*, January 19, 2002.

Cohen, Avner. *Israel and the Bomb*. New York: Columbia University Press, 1998.

Cordesman, Anthony H. "The Changing Military Balance in the Gulf." *Middle East Policy* 6, no. 1 (June 1998): 25–44.

Cottrell, Alvin J, ed. *The Persian Gulf States: A General Survey*. Baltimore: Johns Hopkins University Press, 1980.

Crossette, Barbara. "Iraq Suspected of Secret Germ War Effort." *New York Times*, February 8, 2000.

———. "Pressing for Iraqi's Overthrow, U.S. Appeals for Arab Support." *New York Times*, December 9, 1998.

Davies, Peter. "Climate Change: What If We Can't Unravel the Gordian Knot?" *Energy Economist*, no. 225 (July 2000): 9–14.

Deese, David A., and Joseph S. Nye. *Energy and Security*. Cambridge, Mass.: Ballinger, 1981.

Demirmen, Ferruh. "Despite Recent Gains in Momentum, Prospects for the Baku-Ceyhan Caspian Oil Export Line Remain Doubtful." *Oil and Gas Journal* 97, no. 46 (November 15, 1999): 23–28.

Diamond, Larry. "Is the Third Wave Over?" *Journal of Democracy* 7, no. 3 (July 1996): 20–37.

Dobbs, Michael. "Old Strategy on Iraq Sparks New Debate." *Washington Post*, December 27, 2001.
Drozdiak, William. "Iraq Imperils OPEC Plan." *Washington Post*, February 10, 2001.
Dunn, Michael C. "The Kurdish Question: Is There an Answer? A Historical Overview." *Middle East Policy* 4, no. 1 (September 1995): 72–86.
Easton, Nina J. "The Hawk." *Washington Post*, December 27, 2001.
Economist Intelligence Unit. *Country Analysis: Iran*. London, July 2001, on-line at www.eiu.com.
"Editorial: Oil Prices and Layoffs." *Oil and Gas Journal* 97, no. 13 (March 29, 1999): 13.
Ehteshami, Anoushiravan. *After Khomeini: The Iranian Second Republic*. London: Routledge, 1995.
Emerson, Sarah. "SPR Drawdowns Trigger Law of Unintended Consequences." *Oil and Gas Journal* 99, no. 50 (December 10, 2001): 24–30.
Energy Information Administration (EIA). (Unless specified otherwise, publications of the EIA are on-line at www.eia.doe.gov.)
———. *Annual Energy Outlook*. Washington, D.C.: Government Printing Office, 2001.
———. *Caspian Regional Conflict*, June 2000.
———. *Caspian Tables and Maps*, July 2000.
———. *Country Profile: Azerbaijan*, May 2001.
———. *Country Profile: China*, April 2001.
———. *Country Profile: Iran*, May 2001.
———. *Country Profile: Iraq*, 2001.
———. *Country Profile, Japan*, April 2001.
———. *Country Profile: Kazakhstan*, May 2001.
———. *Country Profile: Russia*, October 2001.
———. *Country Profile: Saudi Arabia*, November 2000.
———. *Country Profile: USA*, April 2001.
———. *Country Profile: Turkey*, July 2001.
———. *East Asia: The Energy Situation*, June 2000.
———. *Energy in the United States: A Brief History and Current Trends*. Washington, D.C.: Government Printing Office, 2000.
———. *Impacts of the Kyoto Protocol on U.S. Energy Markets and Economic Activity*, 2000.
———. *International Energy Outlook, 2001*. Washington, D.C.: Government Printing Office, 2001.
———. *International Petroleum Monthly*. Washington, D.C.: Government Printing Office, 2001.
———. *North Sea*, February 2001.
———. *Oil and Gas Export Pipelines*, October 2001.
———. *OPEC Revenues Fact Sheet*, various months and years.
———. *Persian Gulf Oil and Gas Exports Fact Sheet*, February 2001.

———. *Potential Oil Production from the Coastal Plain of the Arctic National Wildlife Refuge: Updated Assessment,* May 2000.

———. *Russia: Energy Sector Restructuring,* October 2001.

———. *Twenty-fifth Anniversary of the 1973 Oil Embargo.* Washington, D.C.: Government Printing Office, 1998.

———. *U.S. Natural Gas Markets: Recent Trends and Prospects for the Future,* May 2001.

———. *World Energy Sanctions,* December 2000.

"Energy Market Report." *Energy Economist* 194 (December 1997): 25–31.

Engur, Emre. "Turkey Determined to Remain at the Center of East-West Energy Corridor." *Oil and Gas Journal* 100, no. 2 (January 14, 2000): 60–64.

"Enterprise Leads Way in South Pars." *Middle East Economic Digest* 45, no. 1 (January 5, 2001): 8.

European Commission. *Green Paper: Towards a European Strategy for the Security of Energy Supply.* Brussels: European Commission, 2000.

Farouk-Sluglett, Marion, and Peter Sluglett. *Iraq since 1958: From Revolution to Dictatorship.* London: KPI, 1987.

Feldman, Shai. "The Bombing of Osiraq—Revisited." *International Security* 7, no. 2 (fall 1982): 114–42.

Fesharaki, Fereidun, and Mehdi Varzi. "Investment Opportunities Starting to Open Up in Iran's Petroleum Sector." *Oil and Gas Journal* 98, no. 7 (February 14, 2000): 44–52.

Forsythe, Rosemarie. *The Politics of Oil in the Caucasus and Central Asia.* London: Oxford University Press, 1996.

Frantz, Douglas. "Russia's New Reach: Gas Pipeline to Turkey." *New York Times,* June 8, 2001.

Gaddy, Dean E. "Iran Expands Middle East Influence." *Oil and Gas Journal* 99, no. 10 (March 5, 2001): 74–81.

"Gas Exports to Rise as New Lines Come on Stream." *Petroleum Economist* 67, no. 1 (January 2000): 40.

Gavin, James. "The Walls Come Down." *Middle East Economic Digest* 44, no. 20 (May 19, 2000): 4.

Gellman, Barton. "Annan Suspicious of UNSCOM Role." *Washington Post,* January 6, 1999.

———. "Israel Gave Key Help to UN Team in Iraq." *Washington Post,* September 29, 1998.

Ghanem, Shokri, Rezki Lounnas, and Garry Brennand. "The Impact of Emissions Trading on OPEC," *OPEC Review* 23, no. 2 (June 1999): 79–112.

Ghanem, Shokri, Rezki Lounnas, D. Ghasemzadeh, and Garry Brennand. "Oil and Energy Outlook to 2020: Implications of the Kyoto Protocol." *OPEC Review* 22, no. 2 (June 1998): 64–83.

Ghoreishi, Ahmad, and Dariush Zahedi. "Prospects for Regime Change in Iran." *Middle East Policy* 5, no. 1 (January 1997): 85–101.

Gigot, Paul A. "A Great American Screw-Up: The U.S. and Iraq, 1980–1990." *National Interest* 22, no. 1 (winter 1990/91): 3–10.

Gokay, Bulent, ed. *The Politics of Caspian Oil*. New York: Palgrave, 2001.

Gorst, Isabel. "Change Looms over the Mighty Gazprom," *Petroleum Economist* 67, no. 9 (September 2000): 16–20.

Graham-Brown, Sarah. "Sanctioning Iraq: A Failed Policy." *Middle East Report* 215 (Summer 2000): 8–35.

"Great Expectations." *Middle East Economic Digest* 44, no. 37 (September 15, 2000): 25–44.

Gurney, Judith. "U.S. Faces Natural Gas Price Shock," *Energy Economist*, no. 229 (November 2000): 15–18.

Halliday, Denis. "The Impact of the UN Sanctions on the People of Iraq." *Journal of Palestine Studies* 28, no. 2 (winter 1999): 29–37.

———. "Iraq and the UN's Weapons of Mass Destruction." *Current History* 98, no. 625 (February 1999): 65–68.

Halliday, Fred. "Condemned to React, Unable to Influence: Iran and Transcaucasia." In *Transcaucasian Boundaries,* ed. John F. R. Wright, Suzanne Goldenberg and Richard Schofield, 71–88. London: UCL Press, 1996.

Hashim, Ahmed. *The Crisis of the Iranian State*. New York: Oxford University Press, 1995.

———. "Iraq: Profile of a Nuclear Addict." *Brown Journal of World Affairs* 4, no. 1 (January 1997): 103–26.

Hersh, Seymour M. "Analysis of National Security: The Iran Game." *The New Yorker*, December 21, 2001, on-line at www.newyorker.com.

Herzig, Edmund. "Iran and Central Asia." In *Central Asian Security: The New International Context*, ed. Roy Allison and Lena Jonson, 171–98. Washington, D.C.: Brookings Institution, 2001.

Heywood, Neil. "Farewell to Self-Sufficiency," *Petroleum Economist* 67, no. 12 (December 2000): 3–5.

Hochstrader, Lee. "Iran Implicated by Captain in Seized Weapons Shipment." *Washington Post*, January 9, 2002.

Horsnell, Paul. "Caspian Oil and Gas: A Game, If Not a Great Game." Oxford Institute for Energy Studies, *Monthly Comment*, January 1999, on-line at www.oxfordenergy.org.

Hulse, Carl. "U.S. Said to Approve Anti-Hussein Funding." *New York Times*, February 2, 2001.

Hunter, Shireen T. *The Transcaucasus in Transition: Nation- Building and Conflict*. Washington, D.C.: Center for Strategic and International Studies, 1994.

Huntington, Samuel. "After Twenty Years: The Future of the Third Wave." *Journal of Democracy* 8, no. 4 (October 1997): 4–12.

Ibrahim, Ibrahim, *The Gulf Crisis: Background and Consequences*. Washington: Center for Contemporary Arab Studies, Georgetown University, 1992.

Ibrahim, Youssef M. "Higher Hopes in Baghdad for Ending UN Embargo." *New York Times*, October 18, 1998.

International Emissions Trading Association. *What Is the Science behind Climate Change?* August 2001, on-line at http://ieta.org.

International Energy Agency. *Caspian Oil and Gas*. Paris: Organization for Economic Cooperation and Development, 1998.

International Institute for Strategic Studies. *Strategic Survey*. London: Oxford University Press, 1995.

International Monetary Fund. *World Economic Outlook*. New York: Oxford University Press, 1996.

"Interview with President Khatami." *New York Times*, November 10, 2001.

"Iran Offers Discounts to Support Caspian Export Offer." *Middle East Economic Digest* 45, no. 1 (January 5, 2001): 9.

"Iran Reveals Second Big Gas Discovery." *Middle East Economic Digest* 44, no. 34 (August 12, 2000): 12.

"Iran's Expediency Council Rescues Buybacks as Contracts Spending Rolls Over into New Budget Year." *Middle East Economic Survey* 44, no. 8 (February 19, 2001), on-line at www.mees.com.

"Iran's NIOC Discovers Supergiant Gas, Condensate Field." *Oil and Gas Journal* 98, no. 18 (May 1, 2000): 40.

"Iraq: Russia Gets $40 Billion in Contracts." *Moscow Times*, October 1, 2001, on-line at www.themoscowtimes.com.

"Iraq: Sanctions Policy Reaches Fork in the Road." *Petroleum Economist* 66, no. 2 (February 1999): 39–40.

"Iraq Changes Terms of Future Oil Project Contracts." *Middle East Economic Digest* 44, no. 29 (July 21, 2001): 10.

Jonas, Theodore C. "Parting the Sea: Caspian Littoral States Seek Boundary Disputes' Resolution." *Oil and Gas Journal* 99, no. 22 (May 28, 2001): 66–69.

Jones, Peter E. *Oil: A Practical Guide to the Economics of World Petroleum*. New York: Nichols, 1988.

Karabell, Zachary. "Backfire: U.S. Policy toward Iraq, 1988–2 August 1990." *Middle East Journal* 49, no. 1 (winter 1995): 28–47.

Karlsson, Svante. *Oil and the World Order: American Foreign Oil Policy*. Totowa, N.J.: Barnes and Noble Books, 1986.

Karshenas, Massoud, and Hashem M. Pesaran. "Economic Reform and the Reconstruction of the Iranian Economy." *Middle East Journal* 49, no. 1 (winter 1995): 92–105.

Kay, David A. "Denial and Deception Practices of WMD Proliferators: Iraq and Beyond." *Washington Quarterly* 18, no. 1 (January 1994): 85–105.

Kazemi, Farhad. "Models of Iranian Politics: The Road to the Islamic Revolution and the Challenge of Civil Society." *World Politics* 47, no. 4 (July 1995): 555–74.

Keinon, Herb. "Iran Using PA as Proxy against Israel-Senior Official." *Jerusalem Post*, January 13, 2002.

Kemp, Geoffrey, ed. "Iran: Can the United States Do a Deal?" *Washington Quarterly* 24, no. 1 (winter 1999): 511–23.

———, ed. *Iran's Nuclear Weapons Options: Issues and Analysis*. Washington, D.C.: Nixon Center, 2001.

Kemp, Geoffrey, and Robert E. Harkavy. *Strategic Geography and the Changing Middle East*. Washington, D.C.: Brookings Institution Press, 1997.

Kemp, Geoffrey, and Janice G. Stein, eds. *Powder Keg in the Middle East*. Lanham, Md.: Rowman and Littlefield, 1995.

Kemper, Ria. "EU Looks to Secure Russian Supply Future," *Petroleum Economist* 67, no. 12 (December 2000): 28–29.

Kendell, James M. *Measures of Oil Import Dependence*. Washington, D.C.: Government Printing Office, 2000.

Khripunov, Igor, and Mary M. Matthews. "Russia's Oil and Gas Interest Group and Its Foreign Policy Agenda," *Problems of Post-Communism* 43, no. 3 (May/June 1996): 38–48.

Kissinger, Henry. "Phase II and Iraq." *Washington Post*, January 13, 2002.

"Kissinger on Oil, Food, and Trade." *Business Week*, January 13, 1975, 66–76.

Korchagina, Valeria. "Russia Gears Up to Export More Oil with BPS," *Moscow Times*, November 30, 2001, on-line at www.themoscowtimes.com.

"The Kyoto Protocol, and the U.S. Came, Too," *Energy Economist*, no. 202 (August 1998): 1–9.

Lelyveld, Michael. "Russian, Iranian Presidents Fail to Agree on Caspian Sea Division." *Radio Free Europe*, March 13, 2001, on-line at www.rferl.org.

Lenczowski, George. *American Presidents and the Middle East*. Durham, N.C.: Duke University Press, 1990.

———. "Major Pipelines in the Middle East: Problems and Prospects." *Middle East Journal* 49, no. 2 (spring 1995): 40–46.

———. *Oil and State in the Middle East*. New York: Cornell University Press, 1960.

Lesch, David W., ed. *The Middle East and the United States: A Historical and Political Reassessment*. Boulder, Colo.: Westview Press, 1996.

Longrigg, Stephen H. *Oil in the Middle East*. New York: Oxford University Press, 1968.

Lynch, Colum. "U.N. Chief Faults U.S., Britain for Iraqi Supply Delays." *Washington Post*, March 14, 2000.

Lynch, Colum, and John Lancaster. "UN Votes to Renew Iraq Inspections." *Washington Post*, December 17, 1999.

Lynch, Michael C. "Oil Prices Enter a New Era." *Oil and Gas Journal* 99, no. 7 (February 12, 2001): 20–30.

McCutcheon, Hilary. "Caspian Production Potential: Discoveries Alter Caspian

Region Energy Potential," *Oil and Gas Journal* 99, no. 51 (December 17, 2001): 18–25.

McDonald, Paul. "What Is the Price of Oil?" *Energy Economist* 229 (November 2000): 5–7.

McDowall, Angus. "Special Report Oil and Gas: Taking Stock," *Middle East Economic Digest* 45, no. 42 (October 19, 2001): 23–38.

McGovern, George. "The Future Role of the U.S. in the Middle East." *Middle East Policy* 1, no. 3 (September 1992): 3–15.

McLachlan, Keith. "Oil in the Persian Gulf Area." In *The Persian Gulf States: A General Survey,* ed. Alvin J. Cottrell, 195–224. Baltimore: Johns Hopkins University Press, 1980.

McNaugher, Thomas L. "Arms Sales and Arms Embargoes in the Persian Gulf: The Real Dilemmas of Dual Containment." In *Powder Keg in the Middle East,* ed. Geoffrey Kemp and Janice Gross Stein, 337–60. Lanham, Md.: Rowman and Littlefield, 1995.

Malone, David. "Good-bye, UNSCOM: A Sorry Tale in U.S.–UN Relations." *Security Dialogue* 30, no. 4 (December 1999): 393–413.

Manning, Robert A. "The Asian Energy Predicament," *Survival* 42, no. 3 (spring 2000): 73–88.

Manning, Tom. "Asian Financial Crisis to Slow Growth in Global Oil Demand." *Oil and Gas Journal* 96, no. 21 (May 4, 1998): 41–44.

Marlowe, John. *The Persian Gulf in the Twentieth Century.* New York: Praeger, 1962.

Merzliakov, Iurii. "Legal Status of the Caspian Sea." *International Affairs* 45, no. 9 (March 1999): 33–39.

Miller, Judith. "UN Panel Urges New Inspections in Iraq." *New York Times,* March 28, 1999.

Moore, Molly. "Mistrust Still Divides Iran and U.S." *Washington Post,* June 22, 2001.

"More Trouble from Saddam." *Christian Science Monitor,* March 9, 1995.

Morris, Harvey. "Israel Fears Threat of Iran Links with Palestinians." *Financial Times,* January 15, 2002.

Morse, Edward L. "A New Political Economy of Oil?" *Journal of International Affairs* 53, no. 1 (fall 1999): 1–29.

———. "The Tigers Are Roaring Back . . . but Their Stripes Have Changed," *Middle East Economic Survey* 42, no. 39 (September 27, 1999), on-line at www.mees.com.

Morse, Edward L., and Amy M. Jaffe. *Strategic Energy Policy Challenges for the Twenty-first Century.* Houston: Rice University, Baker Institute for Public Policy, 2001.

al-Naimi, Ali. "Saudi Oil Policy Combines Stability with Strength, Looks for Diversity." *Oil and Gas Journal* 98, no. 3 (January 17, 2000): 16–18.

Namazi, Siamak. "U.S.–Iran Relations after 11 September." *Payvand's Iran News*, October 8, 2001, on-line at www.payvand.com.

Nanay, Julia. "The U.S. in the Caspian: The Divergence of Political and Commercial Interests." *Middle East Policy* 6, no. 2 (October 1998): 150–57.

National Energy Policy Development Group. *National Energy Policy*. Washington, D.C.: Government Printing Office, 2000.

"Negotiations for Saudi Natural Gas Initiative to Start in Second Half 2000." *Middle East Economic Survey* 43, no. 19 (May 8, 2000), on-line at www.mees.com.

"No Inflation Fear." *Petroleum Economist* 67, no. 1 (January 2000): 38.

Nonneman, Gerd. "Saudi-European Relations, 1902–2001: A Pragmatic Quest for Relative Autonomy." *International Affairs* 77, no. 3 (July 2001): 631–61.

"OPEC Ministers Approve Huge 1.7 Million B/D Production Cuts." *Middle East Monitor* 29, no. 3 (March 1999): 22.

Oren, Amir. "Israel Asks America to Strike Western Iraq First, If It Decides to Fight Saddam." *Ha'aretz*, January 2, 2002.

O'Sullivan, Meghan L. *Iraq: Time for a Modified Approach*. Washington, D.C.: Brookings Institution, 2001.

Ottaway, David B. "U.S. Oil CEOs and Saudis to Meet." *Washington Post*, April 13, 2000.

Ottaway, David B., and Martha M. Hamilton. "Saudis Talk with Seven U.S. Oil Firms." *Washington Post*, September 30, 1998.

Pahlavi, Muhammad R. *Answer to History*. New York: Stein and Day, 1980.

Pala, Christopher. "AGIP of Italy Picked to Manage Big Kazakh Oil Field." *New York Times*, February 13, 2001.

"Palestinian-Iran Ties Worry Israel." Associated Press, January 8, 2002.

Petrossian, Vahe. "Raising the Stakes." *Middle East Economic Digest* 44, no. 36 (September 8, 2000): 4–5.

———. "Special Report Gas: Aiming East." *Middle East Economic Digest* 44, no. 31 (August 4, 2000): 23–28.

———. "Special Report Iran: Year of Decision." *Middle East Economic Digest* 45, no. 5 (February 2, 2001): 23–28.

Pianin, Eric. "U.S. Faces an Energy Shortfall, Bush Says." *Washington Post*, March 20, 2001.

"Producers Agree to Maintain Production Cutbacks until End-March; Positions Differ on What to Do Next." *Middle East Economic Survey* 43, no. 5 (January 31, 2000), on-line at www.mees.com.

"Rafsanjani Says U.S. Aiming to 'Besiege' Iran, Warns Washington, Europe." *Iran Mania*, October 10, 2001, on-line at www.iranmania.com.

Ramazani, R. K. "Iran and the Arab-Israeli Conflict." *Middle East Journal* 32, no. 4 (fall 1978): 413–28.

Richards, Alan. "Oil Wealth in the Arab World: Whence, to Whom, and Whither?"

In *The Arab World Today*, ed. Dan Tschirgi, 67–77. Boulder, Colo.: Rienner, 1994.

Rodenbeck, Max. "Is Islamism Losing Its Thunder?" *Washington Quarterly* 21, no. 2 (spring 1998): 177–93.

"Russian Moves Ahead with Caspian Petroleum Projects." *Middle East Economic Survey* 43, no. 17 (April 24, 2000), on-line at www.mees.com.

Sadowski, James Y. "Prospects for Democracy in the Middle East: The Case of Kuwait." *Fletcher Forum of World Affairs* 21, no. 1 (winter/spring 1997): 57–72.

Saivetz, Carol R. "Caspian Geopolitics: The View from Moscow." *Brown Journal of World Affairs* 7, no. 2 (summer/fall 2000): 53–61.

"Saudi Arabia on Kyoto Protocol Developments," *Middle East Economic Survey* 44, no. 14 (April 2, 2001), on-line at www.mees.com.

"Saudi Arabia Seizes IPSA Pipeline." *Middle East Economic Survey* 44, no. 25 (June 18, 2001), on-line at www.mees.com.

"Saudi Warning of Further Export Hike Trims $2/B Off Crude Prices." *Middle East Economic Survey* 43, no. 28 (July 10, 2000), on-line at www.mees.com.

Schneider, Howard. "Bombing in Saudi City Kills American." *Washington Post*, October 7, 2001.

———. "Review of Iraq Sanctions Reflects State of Disarray." *Washington Post*, March 3, 2001.

Sciolino, Elaine. "Iranians Back Their Leader's Volatile Experiment." *New York Times*, June 10, 2001.

"Security Council Renews Oil-for-Food Program, Pledges to Revise Iraqi Sanctions Next June." Middle East Economic Survey 44, no. 49 (December 3, 2001), on-line at www.mees.com.

Sedan, Gil. "Iran Has Long Anti-Israel History." *Cleveland Jewish News*, January 9, 2002, on-line at www.clevelandjewishnews.com.

Seymour, Ian. "Kuwait's Upstream Oil Opening in the Context of Parallel Development Elsewhere." *Middle East Economic Survey* 43, no. 1 (January 3, 2000), on-line at www.mees.com.

———. "Opportunities for Upstream Investment in the Middle East by IOCs." *Middle East Economic Survey* 43, no. 24 (June 12, 2000), on-line at www.mees.com.

Shaffer, Brenda. *U.S.–Russian Relations: Implications for the Caspian Region*. Cambridge: Harvard University Press, 2001.

Shaviv, Miriam. "A Chill Wind from Tehran." *Jerusalem Post*, January 20, 2002.

Shelley, Toby. "BG in LNG Venture with Iran." *Financial Times*, September 26, 2000.

Shenon, Philip. "U.S. General Warns of Dangers in Trying to Topple Iraqi." *New York Times*, January 29, 1999.

Sicker, Martin. *The Bear and the Lion*. New York: Praeger, 1988.

Sinker, Roland. "The Management of a Transboundary Energy Resource: The Oil and Gas of the Caspian Sea." In *The Politics of Caspian Oil,* ed. Bulent Gokay, 51–109. New York: Palgrave, 2001.

Sipress, Alan. "More Iraqi Oil Evading Sanctions." *Washington Post,* February 18, 2001.

Smith, Jeffrey. "Khatami Wants to End Terrorism, Officials Say." *Washington Post,* May 9, 1998.

Smith, Robert. "Politics, Production Levels to Determine Caspian Area Energy Export Options." *Oil and Gas Journal* 99, no. 22 (May 28, 2001): 33–38.

Snyder, Jed C. "The Road to Osiraq: Baghdad's Quest for the Bomb." *Middle East Journal* 37, no. 4 (fall 1983): 565–93.

Sobhani, Sohrab. *The Pragmatic Entente.* New York: Praeger, 1989.

Soligo, Ronald, and Amy Jaffe. *The Economics of Pipeline Routes: The Conundrum of Oil Exports from the Caspian Basin.* Houston: Rice University Press, 1998.

Stevens, Paul. "Consumer Governments, Energy Security of Supply, and the Aftermath of 11 September." *Middle East Economic Survey* 44, no. 48 (November 26, 2001), on-line at www.mees.com.

——. *The Economics of Energy.* Cheltenham: Edward Elgar, 2000.

Stocking, George W. *Middle East Oil: A Study in Political and Economic Controversy.* Nashville: Vanderbilt University Press, 1970.

Stork, Joe. *Middle East Oil and the Energy Crisis.* New York: Monthly Review Press, 1975.

"Studies Start for IGAT-4." *Middle East Economic Digest* 45, no. 31 (August 3, 2001): 8.

Takin, Manouchehr. "Iranian Gas to Europe?" *Middle East Economic Survey* 42, no. 15 (April 12, 1999), on-line at www.mees.com.

"Tehran Seeks More Gas Buy-Backs." *Middle East Economic Digest* 44, no. 35 (September 1, 2000): 11.

Telhami, Shibley. "Time for Realism on Handling Iraq." *Washington Post,* June 20, 2001.

"Ten Leading Lawmakers Urge Targeting of Iraq." *Washington Post,* December 6, 2001.

Tillman, Seth P. *The United States in the Middle East.* Bloomington: Indiana University Press, 1982.

"TotalFinaElf Invites Bids for Doroud Onshore Package in Iran." *Middle East Economic Digest* 44, no. 31 (August 4, 2000): 11.

Troner, Alan. "Russian Far East Natural Gas." *Oil and Gas Journal* 99, no. 10 (March 5, 2001): 68–72.

Tschirgi, Dan, ed. *The Arab World Today.* Boulder, Colo.: Rienner, 1994.

United Nations Development Program. *Human Development Report.* New York: Oxford University Press, 1999.

"U.S. Crude Reserves Plunged 7 Percent in 1998." *Oil and Gas Journal* 98, no. 1 (January 3, 2000): 32.

"U.S. Reviews Iran and Sanctions Policies." *Middle East Economic Digest* 45, no. 5 (February 2, 2001): 3.

"U.S. Sanctions: Fresh Thinking on Capitol Hill." *Middle East Economic Digest* 45, no. 14 (April 6, 2001): 29–30.

"U.S./UN Begin Review of Iraqi Sanctions." *Middle East Economic Survey* 44, no. 9 (February 26, 2001), on-line at www.mees.com.

U.S. Defense Department. *Proliferation: Threat and Response.* Washington, D.C.: Government Printing Office, 2001.

U.S. Energy Department. *Strategic Petroleum Reserve.* August 2000. On-line at www.fe.doe.gov/spr/spr.html.

U.S. State Department. *Patterns of Global Terrorism.* Washington, D.C.: Government Printing Office, 2001.

von der Mehden, Fred R. *Japan's Relations with Primary Energy Suppliers.* Baker Institute Study no. 13. Houston: Rice University, May 2000.

Weinbaum, M. G. "Iran and Israel: The Discrete Entente." *Orbis* 18, no. 1 (winter 1975): 1070–87.

Weiner, Tim. "U.S. Spied on Iraq under UN Cover, Officials Now Say." *New York Times,* January 7, 1999.

Weller, Marc. "The U.S., Iraq, and the Use of Force in Unipolar World." *Survival* 41, no. 4 (winter 1999–2000): 81–100.

Whitaker, Brian. "Voyage of the Arms Ship." *Guardian,* January 14, 2002.

Winfield, Nicole. "UN: Euro Account for Iraq Feasible." *Washington Post,* October 28, 2000.

Woolsey, James R. "Appeasement Will Only Encourage Iran." *Survival* 38, no. 4 (winter 1996): 19–21.

World Bank. *World Debt Tables.* New York: Oxford University Press, 1996.

"World News: More Doubts over Azerbaijan Energy." BBC, July 25, 2001, on-line at http://news.bbc.co.uk.

"Worldwide Look at Reserves and Production." *Oil and Gas Journal* 98, no. 51 (December 18, 2000): 122.

Wright, F. R., Suzanne Goldenberg, and Richard Schofield, eds. *Transcaucasian Boundaries.* London: UCL Press, 1996.

Wright, Robin. "Iran's Greatest Political Challenge: Abdol Karim Soroush." *World Policy Journal* 24, no. 2 (summer 1997): 67–74.

Yamani, Ahmed Zaki. "OPEC Should Take Long-Term Approach to Balancing Oil Supply-Demand Equation." *Oil and Gas Journal* 97, no. 38 (September 20, 1999): 23–30.

Yaniv, Avner. "Israel Faces Iraq: The Politics of Confrontation." In *Iraq's Road to War,* ed. Amatzia Baram and Barry Rubin, 233–55. New York: St. Martin's Press, 1993.

Young, David. "Politics or Technology." *Oil and Gas Journal* 99, no. 51 (December 17, 2001): 27.

Zacharia, Janine. "World Bank Set to Consider Iran Loans." *Jerusalem Post*, May 10, 2001.

Zakaria, Fareed. "The Rise of Illiberal Democracy." *Foreign Affairs* 76, no. 6 (November/December 1997): 22–43.

Zarif, Muhammad Javad, and Muhammad Reza Alborzi. "Weapons of Mass Destruction in Iran's Security Paradigm: The Case of Chemical Weapons." *Iranian Journal of International Affairs* 11, no. 4 (Winter 1999): 511–23.

Index

Page numbers in *italics* indicate figures and tables.

Abkhazia, 157, 167
"above-the-ground risk," 142–43
Abraham, Spencer, 183n.22
Afghanistan: hydrocarbon reserves of, 154; Iran and, 134–37; pipelines and, 167; Powell on Iraq compared to, 100–101; Taliban and, 114
Agip, 153, 168
Alaska, *3, 7,* 13
Albright, Madeleine, 95
Algeria, 28, 61, 179n.9
Algiers Accord, 131
Aliyev, Heydar, 158, 159, 160, 168
al-Qaeda, 135–36
Alviri, Morteza, 135
Anderson, Terry, 130
Anglo-Iranian Oil Company, 106–7
Anglo-Persian Oil Company, 105–6, 182n.2
Annan, Kofi, 78, 92, 93
Anti-terrorism and Effective Death Penalty Act of 1996, 129–31
al-Aqsa intifada, 70, 82
Arabian American Oil Company (Aramco), 45, 61, 183n.39
Arab-Israeli conflict, 69–70, 72
Arab League, 183n.18
Arafat, Yasser, 124, 125
Arbusto Energy, Inc., 1
Arctic National Wildlife Refuge, Alaska, *3,* 13
Armenia, 158–60
Armitage, Richard L., 100
arms sales: Iran to Palestinian Authority, 125; to Saudi Arabia, 49; by Turkmenistan, 153; al-Yamamah deal, 29–30

Ashcroft, John, 128
assessment of resources of Caspian states, 143–46
Automobile Efficiency Standards, 8
Azerbaijan: Caspian Sea and, 140, *142–43,* 161, 164; ethnic conflict in, 158–60; Iran and, 153, 165, 171–72; Russia and, 150, 162
Azerbaijan International Operating Company (AIOC), 142, 161, 168
al-Aziz, Abdullah Ibn Abd (crown prince), 61, 64, 65, 69
al-Aziz, Saud Ibn Abd (king), 43

Baghdad Pact, 72, 108
Bahrain, 3
Baku-Ceyhan pipeline, 167–68
Baku region, 140
Bechtel, 168
Ben-Gurion, David, 124
BG International, 115, 153, 168
Biological Weapons Convention, 132
Blue Stream, 169–70
Bosphorus Strait, 166
BP/Amoco, *144, 145,* 165, 168–69
Brazil, 20
Britain. *See* United Kingdom
British Petroleum, 106. *See also* BP/Amoco
Bush, George, 17, 55, 75, 96
Bush, George W.: energy security and, 1; Hemispheric energy policy and, 18–19; Iraq and, 97; Kyoto Protocol and, 41, 67; nuclear power and, 15–16; payments to victims of terrorism and, 130–31; sanctions against Iran, 121–22; Strategic Petroleum Reserve and, 17; on terrorism, 70

"buy-back" model, 117–18
"B-Wave," 47

California-Arabian Standard Oil Company, 43, 45
Canada, 11, 18–19
carbon sink, 40
Carter, Jimmy, 12, 49, 68, 131
Caspian Oil Company, 150
Caspian Pipeline Consortium (CPC), 170
Caspian region: assessment of resources of, 143–46; ethnic divisions in, 155–60; geopolitical considerations in, 147–55; overview of, 140, 142–43; pipeline diplomacy, 166–72; September 11 and, 154–55
Caspian Sea: legal status of, 160–66; map of, *141*; outlook for, 172–73
Chalabi, Fadhil J., 86
Chavez, Hugo, 56
Chechnya, 152, 156, 167
Chemical Weapons Convention, 132
Cheney, Dick, 1, 122, 184n.55
Chevron: Caspian region and, 149; Iran and, 121; Kazakhstan and, 142, 170; Saudi Arabia and, 43, 61
ChevronTexaco, 149
China, 32, 35–37, 133
Clinton, Bill: al-Aqsa intifada and, 82; Baku-Ceyhan pipeline and, 168; Iraq and, 97; Iraqi Liberation Act, 96; Khobar Towers bombing and, 128, 129; Kyoto Protocol and, 41; payments to victims of terrorism and, 130; public lands and, 13; sanctions against Iran, 118–19, 120, 121; Strategic Petroleum Reserve and, 17, 58
coal, 4–5
Collins, Susan, 58
competition: in Caspian region, 147–48, 173; for Caspian resources, 155; for foreign capital, 62–63
Conference on Disarmament, 132
Conoco, 119, 121
conservation, 14–15, 57
Consortium, 107

Dagestan, 156, 167
D'Arcy, William Knox, 105

deep-water drilling technology, 14
demand for energy, 2
Demirel, Suleiman, 168
Department of Energy, 12, 58, *144*
directional drilling, 14
domestic oil industry, 55, 57
Downing, Wayne A., 99
Dubai Fateh crude, 47

Economic Cooperation Organization, 147, 152
economy: of Azerbaijan, Kazakhstan, and Turkmenistan, 142; energy efficiency and, 15; of Iran, 109; of Iraq, 81–82; of Pacific Asia region, 34–35, 55; of Saudi Arabia, 63, 68–69; of United States, 2, 53–54. *See also* price of oil
Egypt, 50, 70
Eisenhower, Dwight, 11, 17
Elchibey, Abulfaz, 158
electric power plants and natural gas, 5
Elf Aquitaine, 115
Emergency Highway Conservation Act, 11
Energy Charter Treaty, 31–32
energy dependence compared to vulnerability, 9
energy intensity, 15
energy policy: of G. W. Bush administration, 13–14; of Carter administration, 12; of Eisenhower administration, 11; of Ford administration, 12; hemispheric, 18–19, 24; interdependence, 9, 23, 26; lack of, 1–2; liberalization of, 2; in 1980s and 1990s, 12; of Nixon administration, 11–12; proposals for, 2–3, 22–23; restraints on, 14; of F. D. Roosevelt administration, 9–10; since 1950s, 10–11
Energy Policy and Conservation Act, 12, 17
energy prices and recessions, 2, 53–54
Eni S.p.A., 115, 169
Enterprise, 115
ethnic divisions in Caspian region, 155–60
European Union: as energy consumer, 26–27, *27*; Gulf Cooperation Council and, 29–30; Iran and, 29; Iran-Libya Sanctions Act Extension and, 122; Kyoto Protocol and, 40; members of, 26; natural gas and, 27–28; as net energy importer, 28–29; oil and, 27; Russia and,

31–32
executive orders, 120–21
Exxon, 61
ExxonMobil: al-Ghawar field and, 64–65; Iran and, 121; Oguz field and, 145; Russia and, 181n.18; Tengizchevroil and, 170

Fahd (king), 64, 70
al-Faisal, Saud (prince), 64
Flatow, Stephen M., 129
Ford, Gerald, 12, 17
foreign investment: in Azerbaijan, Kazakhstan, and Turkmenistan, 142–43; competition for, 62–63; in Iran, 61, 63, 108–10; in Iraq, 61, 63, 82, 85–87; Persian Gulf investors and, 21–22; in Russia, 31, 181n.18; in Saudi Arabia, 22, 61–66
foreign policy: role of energy in, 2; Russian, 33
fossil fuels: in Pacific Asia, 34, 35; in United States and European Union, 27; U.S. consumption of, 4, 5. See also coal; natural gas; oil
Fox, Vicente, 19
France: Azerbaijan, Armenia, and, 159; Iraq and, 29, 86, 92, 96
Freedom for Russia and Emerging Eurasian Democracies and Open Markets Support Act of 1992, 159
Friendship Treaty of 1921, 160–61
fuel efficiency, 8
fuel mix: coal, 4–5; diversification of, 4; natural gas, 5–6; oil, 6–9; overview of, 3

Gaz de France, 32
Gazprom, 31, 32, 120, 150, 169
General Electric Capital, 168
geopolitical issues in Caspian area: Iran and, 151–54; overview of, 147–48; Russia and, 149–51; Turkey and, 151; United States and, 148–49
Georgia, 156–57
globalization: of energy scene, 24, 26; of oil markets, 9
global warming, 39–42
greenhouse gases, 181n.33
Gulf Cooperation Council, 29–30
Gulf of Mexico, 13–14
Gulf Oil Corporation, 182n.2
Gulf war, 74–75
Gunashli, 142
GUUAM Group, 147, 157

al-Hakim, Muhammad Baqir, 98
Halevy, Ephraim, 126
Halliburton Company, 1, 122
Halliday, Denis, 81
Hemispheric energy policy, 18–19, 24
Higgins, William R., 130
Hizbollah, 124, 125, 129, 189n.50
Homa, 111
Hussein, Saddam: ethnic conflict and, 96; oil markets and, 85; post–Gulf war, 74, 75, 101; sanctions and, 82; weapons of, 89, 91

Ickes, Harold, 17
Independent Petroleum Association of America, 10–11
India, 113–15
Indonesia, 61
in-kind payments to Strategic Petroleum Reserve, 17
interdependence, 9, 23, 26
International Atomic Energy Agency, 133
International Energy Agency, 49–50
International Energy Forum, 67
International Institute for Strategic Studies, 143
Iran: Afghanistan and, 134–37; Azerbaijan and, 153, 165, 171–72; Azeri citizens of, 157–58; "buy-back" model, 117–18; Caspian area and, 144–45, 151–54, 163–64, 170–72; description of, 103; economic sanctions against, 118–23; energy sector in, 103, 105–8; European Union and, 29; foreign investment in, 61, 63, 108–10; global energy market and, 3; India and, 113–15; Iraq and, 90; Israel and, 123–26; Japan and, 38, 116; Kazakhstan and, 153, 171–72; map of, 104; natural gas in, 110–15; oil concessions, 105–6; oil field development deals of, 115–17; oil production, 109–10, 110; revolution of 1979, 47, 50, 103, 108, 138; Russia and, 133, 153, 160–

Index

Iran—*continued*
61; sanctions against, 118–23; terrorism and, 126–31; Turkey and, 112–13; Turkmenistan and, 153, 171–72; U.S. interests in, 107–8, 139; weapons of, 131–34. *See also* Iran-Iraq war

Iranian gas trunkline network, 112

Iran-Iraq war, 51, 74, 77

Iran-Libya Sanctions Act, 118, 119–20; extension, 122

Iran Power Plant Management Company, 171

Iraq: G. W. Bush administration and, 87–89; containment strategy, 75–76; creation of, 72; economic sanctions against, 77–78, 80–84, 87–88; embargo on, 78; euro and, 83; foreign investment in, 61, 63, 82, 85–87; France and, 29, 86, 92, 96; global energy market and, 3; invasion of Kuwait, 47, 51, 68, 74; Iran and, 90; Israel and, 90; Kuwait and, 91; map of, 73; no-fly zones, 96; oil for food program, 78, 81; oil industry, 76–80, 84–85; oil policy, 55; oil production, 1970–2000, 77; oil reserves, 60; as oil smuggler, 85; political system in, 97–98, 102; rebellion in, 96; Russia and, 86; Saudi Arabia and, 91, 94, 97; September 11 terrorist attacks and, 99–101; support for opposition in, 95–98; surcharge and, 83; as swing producer, 84–85; Syria and, 83–84, 85; Turkey and, 85, 89, 94, 97; U.S. policy toward, 72, 74, 101–2; weapons programs, 80, 81–82, 89–95. *See also* Iran-Iraq war

Iraqi National Accord, 96, 98

Iraqi National Congress, 96, 98

Iraqi Petroleum Company (IPC), 76–77, 79

Iraq National Oil Company, 77

Islam, 69, 70–71, 152

Islamic Movement of Iraqi Kurdistan, 96

Israel: Iran and, 123–26; Iraq and, 90; oil as political weapon for allies of, 48; Yom Kippur War, 47, 49, 70. *See also* Arab-Israeli conflict

Japan: energy sector in, 37–38; Iran and, 38, 116; Iran-Libya Sanctions Act Extension and, 122; Kyoto Protocol and, 40–41; Russia and, 32–33

Jenco, Lawrence, 130

Jiang Zemin, 32, 37

joint venture, 117

Jordan, 88–89, 98

Karazi, Hamid, 137

Karine A, 125

Karrazi, Kamal, 121, 127, 137

Karroubi, Mehdi, 121

Kashagan block, 145–46

Kay, David, 92

Kazakhstan: Caspian Pipeline Consortium (CPC) and, 170; Caspian Sea and, 140, 142–43, 164–65; Iran and, 153, 171–72; Russia and, 162, 164

Kelberer, John J., 45

Khamenei, Ayatollah, 124, 136

Khan, Ismail, 135

Khatami, Muhammad: Caspian region and, 162; Clinton administration and, 121; Japan, visit to, 116; as moderate, 127; Palestinians and, 125; reforms of, 103, 138; on terrorism, 135–36

Khobar Towers bombing, 128–31

Khomeini, Ayatollah: challenges to regime of, 103; Israel and, 124; political dynamics of Gulf region and, 72; on revolution, 138; Rushdie and, 127

Kissinger, Henry, 49, 99, 100

Kocharian, Robert, 159–60

Kurdish Democratic Party of Iran, 127

Kurdish separatists, 156

Kurdistan Democratic Party, 96

Kurdistan Workers' Party, 156

Kuwait: foreign investment in, 61; global energy market and, 3; Iraq and, 91; Iraqi invasion of, 47, 51, 68, 74

Kyoto Protocol: overview of, 39–40; reactions to, 40–42, 67

Lasmo, 145, 152–53

Libby, I. Lewis, 99

Libya, 61, 102, 118, 119

liquefied natural gas, 112, 179n.9
Lukoil, 145, 150

Mackenzie, Wood, 146
al-Majid, Hussein Kamel, 92
Masood, Ahmad Shah, 135
Master Gas System, 64
Metropolitan Edison Company, 180n.27
Mexico, 11, 18, 19
miles per gallon, 8
Miller, Aleksei, 31
Minsk Group, 158
missile proliferation, 91, 132
"missing barrels" problem, 60
Mobil, 61, 171
Mohtashemi, Ali Akbar, 129
Mossadeq, Muhammad, 107
Movement for a Constitutional Monarchy, 96
Mujahedin-e Khalq (MEK), 90, 128–29
multidependence stragegy, 29
Mutalibov, Ayaz, 158
Mykonos Verdict, 127–28

Nagorno-Karabakh, 158, 167
al-Naimi, Ali, 45, 53, 58
National Energy Policy Development Group, 1, 184n.55
National Iranian Oil Company (NIOC), 106–7
natural gas: demand for, 5–6; European Union and, 27–28; growth of, 5; Iran and, 110–15; North American states and, 18–19; Russia and, 31; Saudi Arabia and, 63–66; Turkmenistan and, 146
Nixon, Richard, 11–12, 49
Non-Proliferation Treaty, 133
Norex, 115
Norsk Hydro, 115
Northern Alliance, 134–35
North Sea, 25, 28
North Sea Brent, 46
Northwestern Europe, 25
Norway, 28
Novorosiisk, 166, 170
nuclear power, 15–16, 37–38

Nuclear Regulatory Commission, 16

Oclean, Abdullah, 156
Offshore Kazakhstan International Operating Company, 145–46, 168
oil: consumption of, 8; dependence on imported, 6–7, 7, 8–9; European Union and, 27; from Persian Gulf, 20–22; significance of, 3; from South Atlantic Region, 20. *See also* pipelines; price of oil
Oil and Gas Journal, 144
Oil Price Protection Act, 58
Oman, 3, 47
Operation Desert Fox, 93
Optimarket, 171
Organization for Economic Cooperation and Development, 49
Organization of Petroleum Exporting Countries (OPEC): creation and function of, 54; oil price volatility and, 51–52, 54–56; price band mechanism, 58; price hikes in 1973, 8; quota discipline and, 60; quota system, 50–51; reference prices, 46
Osiraq nuclear reactor, 90, 92
Ossetia, 157, 167
Oxford Institute for Energy Studies, 143–44

Pacific Asia region: China, 32, 35–37, 133; economy of, 55; Japan, 32–33, 37–38, 40–41, 116, 122; overview of, 34–35, 35
Pakistan, 114
Palestinian Authority, 125
Partnership for Peace Program, 147, 157
Patriotic Union of Kurdistan, 96–97
Patterns of Global Terrorism (U.S. State Department), 126–27
Peres, Shimon, 125
"Peripheral Alliance" strategy, 124
Persian Gulf: importance of oil from, 20–22; imports from as percentage of net oil imports, 21; Pacific Asia region and, 35; U.S. policy in, 75
Petroleos Mexicanos, 19
Petroleum Finance Company, 143

Petroleum Law of 1987, 117
Petronas, 120
pipelines: Azerbaijan, Kazakhstan, Turkmenistan, and, 166–72; between Canada and U.S., 18–19; Central Asian, 154–55, 156–57; Indian-Iranian, 113–15; Iraq and, 79, 83–84; between Russia and Europe, 32
political factors: in Caspian Sea demarcation, 165; Gulf production capacity and, 22; in oil prices, 48–49, 66–67
Powell, Colin: Central Asia and, 154; Iran and, 122, 137; on Iraq, 87, 97; on Iraq compared to Afghanistan, 100–101
price of oil: Caspian exploration and development operations and, 172; effect of shortfall on, 25; effects of high prices, 57–58; history of, 47–51; 1972–2000, 48; outlook for, 59–60; political factors in, 48–49, 66–67; recessions and, 2, 53–54; reference price, 46–47; Saudi Arabia and, 53; shocks, 47, 49–50, 59; stability of, 23; U.S. trade deficit and, 57; volatility since 1997, 51–59
production-sharing agreement, 117
"Project Independence," 11–12
Putin, Vladimir, 30, 31, 150, 162

Qassim, Adb al-Karim, 72
Qatar, 3, 111

Rabbani, Burhanuddin, 135
Rafsanjani, Ali Akbar Hashemi, 111, 127, 136
Rajavi, Massoud, 128
recessions and energy prices, 2, 53–54
reserve to production, 20
Ritter, Scott, 92–93
Roosevelt, Franklin D., 11, 68
Royal Dutch/Shell: Caspian region and, 140, 145; Iran and, 115, 152–53; Saudi Arabia and, 64, 182n.2
Rumsfeld, Donald H., 95–96, 99, 154
Rushdie, Salman, 127
Russia: Armenia and, 159; Azerbaijan and, 150, 162; Blue Stream pipeline, 169–70; Caspian area and, 144–45, 149–51, 162–64; China and, 32; energy sector in, 30–33; European Union and, 31–32; foreign investment in, 31, 181n.18; gas exports to European Union from, 28; Iran and, 133, 153, 160–61; Iraq and, 86; Japan and, 32–33; Kazakhstan and, 162, 164; Turkmenistan and, 164

Sadat, Anwar, 70
sanctions policy: Gulf production capacity and, 22; Iran, 118–23; Iraq, 77–78, 80–84, 87–88
Saudi Arabia: alliance with U.S., 62, 67–68; Arab-Israeli conflict and, 69–70; economic reform in, 68–69; effects of low oil prices on, 55–56; energy policy of, 53–54; foreign investment and, 22, 61–66; gas initiative, 63–66; global energy market and, 3; Iraq and, 91, 94, 97; Iraqi pipelines and, 79; Kyoto Protocol and, 42, 67; map of, 44; militant Islam and, 70–71; nationalization of foreign oil operations, 61; oil explorations in, 43, 45; oil production and share of OPEC total, 52; OPEC and, 52–53, 54; as producer and exporter of oil, 53; September 11 hijackers and, 33; surplus capacity of, 60
Schumer, Charles, 58
security: self-sufficiency compared to, 23; as shared issue, 24, 67
September 11 terrorist attacks: Caspian region policy and, 154–55; dependence on Middle East oil and, 33; energy security and, 22; Iran and, 135–36; Iraq and, 94–95, 99–101; oil demand, prices, and, 59; Saudi Arabia and, 71; United Nations and, 88
Seven Sisters, 182n.2
Shah Deniz, 145, 168–69
Sharon, Ariel, 125
Shatt al-Arab waterway, 75, 90
Shevardnadze, Eduard, 168
South Atlantic Region, 20
South Pars, 111
Standard Oil Company of California, 43
strategic partnerships, 24

Strategic Petroleum Reserve: creation of, 12, 16–17; drawdowns from, 17; mandate of, 58; problems with, 17–18
Straw, Jack, 137
Sudan, 102
supply of energy, 2
Supreme Assembly for the Islamic Revolution in Iraq, 96–97, 98
surplus capacity, erosion of, 60
Sutherland, Thomas, 130
swaps, system of, 58, 170–72
Syria, 83–84, 85, 98, 102

Tabnak, 111
Taliban: Indian-Iranian partnership and, 114; Iran and, 134–35, 136, 137; Unocal and, 155
technical services contract, 117
Tenet, George, 127
Tengizchevroil, 170
Tengiz oil field, 142, 170
terrorism: Iran and, 126–31; war on, 154. *See also* September 11 terrorist attacks
Texaco, 61. *See also* ChevronTexaco
three-dimensional seismic technology, 14
Three Mile Island, 16, 180n.27
Total, 86, 120
TotalFinaElf, 153, 168
trade, international, and oil, 24
Trans-Caspian Pipeline (TCP), 167, 168–69
transportation sector, 8
Treaty of Commerce and Navigation of 1940, 160–61
Trinidad and Tobago, 20
Truman, Harry, 17
Truman Doctrine, 107–8
Turkey: Blue Stream pipeline, 169–70; Caspian area and, 151; Iran and, 112–13; Iraq and, 85, 89, 94, 97; United States and, 149
Turkic nationalism, 157–58
Turkish Petroleum Company, 76
Turkmenistan: Caspian Sea and, 140, 142–43, 165; Gazprom and, 169; Iran and, 153, 171–72; Russia and, 164

United Arab Emirates, 3
United Kingdom: Iran and, 106–7; no-fly zones and, 96; Operation Desert Fox and, 93, 94; smart sanctions program and, 88
United Nations: Convention on the Law of the Sea, 161–62; Framework Convention on Climate Change, 39; Iraq and, 78, 80–81, 88, 92–94
United Nations Special Commission (UNSCOM), 80, 92–94
United States: economy of, 2, 53–54; fossil fuels and, 4, 5, 27; geopolitical issues in Caspian area and, 148–49; Iran and, 107–8, 139; Persian Gulf policy, 75; pipelines between Canada and, 18–19; price of oil and trade deficit in, 57; Saudi Arabia and, 62, 67–68; Turkey and, 149
Unocal Corporation, 154–55
U.S. West Texas Intermediate, 46–47

Van Sponeck, Hans, 81
Venezuela, 20, 56, 61
Vitol, 171
Vyakhirev, Rem, 31

waste, radioactive, 16
weapons: Iran and, 131–34; Iraq and, 80, 81–82, 89–95
Wolfowitz, Paul D., 95–96, 99
Woolsey, James, 99, 100, 126
World Bank, 138
World War I, 72, 106
World War II, 10, 106, 107
World Trade Organization, 69, 138

al-Yamamah deal, 29–30
Yeltsin, Boris, 133
Yemen, 68
Yom Kippur War, 47, 49, 70
Yukos, 150

Zarubezhneft, 86
Zedillo, Ernesto, 19
Zinni, Anthony, 97, 100, 101

Gawdat Bahgat is professor of political science and director of the Center for Middle Eastern Studies at Indiana University of Pennsylvania. He is the author of *The Gulf Monarchies: New Economic and Political Realities* (1997), *The Future of the Gulf* (1997), and *The Persian Gulf at the Dawn of the New Millennium* (1999). He has also published numerous articles on the Persian Gulf and the Caspian Sea in scholarly journals. His work has been translated into Arabic, Russian, German, and Italian.